Scenarios for Technical Communication

Critical Thinking and Writing

Teresa C. Kynell
Northern Michigan University

Wendy Krieg Stone
Interim Technology

Allyn and Bacon

Boston London Toronto Sydney Tokyo Singapore

To Nancy and Joe Hunt & Nedra and Gordon Hall
in Gratitude
and to Mary Ellen Krieg

Vice President, Humanities: Joseph Opiela
Series Editorial Assistant: Rebecca Ritchey
Marketing Manager: Lisa Kimball
Sr. Editorial Production Administrator: Susan McIntyre
Editorial Production Service: Ruttle, Shaw & Wetherill, Inc.
Composition Buyer: Linda Cox
Manufacturing Buyer: Suzanne Lareau
Cover Administrator: Jenny Hart
Electronic Composition: Omegatype Typography, Inc.

A Viacom Company
160 Gould Street
Needham Heights, MA 02494
Internet: www.abacon.com

Library of Congress Cataloging-in-Publication Data

Kynell, Teresa C.
Scenarios for technical communication : critical thinking and writing / Teresa C. Kynell, Wendy Krieg Stone.
p. cm.
Includes index.
ISBN 0–205–27524–9
1. English language—Technical English. 2. Communication of technical information. 3. English language—Rhetoric. 4. Technical writing. 5. Critical thinking. I. Stone, Wendy Krieg. II. Title.
PE1475.K96 1999
808'.0666—dc21 98–35877
CIP

Printed in the United States of America
10 9 8 7 6 5 4 3 2 1 04 03 02 01 00 99 98

Contents

Preface

We first decided to write a textbook based on scenarios because of our firm belief that good technical communicators respond to the unique situations of the workplace, that each day brings writing challenges, that no two technical writers will respond in the same way to the same problem. Most technical communication textbooks, though, include "cases" that are either too advanced for many beginning technical writers or cases that steer the student to one resolution. Scenarios, on the other hand, are created from fictions designed to present the complicated, day-to-day interaction of people involved in workplace decision-making. So the more we investigated the potential of using created scenarios to stimulate both in-class discussions and critical thinking prior to writing, the more we believed they could work in the present educational milieu. The scenarios presented here, however, don't ask students to consider how real events might have been handled differently (like Three Mile Island or the Challenger disasters), nor do these scenarios always provide students with all the answers. We believe that any successful writing project emerges from careful consideration of not only obvious technical matters but also those relationships and often indefinable intangible elements that typically color and problematize a situation. Thus, in these scenarios, students are introduced to a created problem (for example, the location of a proposed soil incineration plant), and might encounter personality conflicts among characters, ethical dilemmas, job-related constraints, and so on. In other words, we want students to deal with the very complex business of thinking—of carefully considering the ramifications of their decisions before they begin the writing process.

We first decided to write a series of scenarios because, simply, we believe that scenarios can provide instructors one of the best means possible to bring real-world situation into the technical communication classroom. Technical communication, as most instructors teach it, is the presentation of rhetorical strategies, models of correct format and writing, and exercises that place students in often unrealistic roles as they attempt to solve arbitrary problems. These arbitrary problems are, in fact, sometimes cases. These cases, though, are typically written so that the student receives just enough information to lead to one, predetermined conclusion. While even these kinds of problems provide some real-world context, they do not place the student in the middle of a complex situation that requires reasoning, critical thinking, and sometimes collaboration to resolve the problem at hand. Scenarios, in many ways, teach students to understand the complexity implicit in workplace writing and how an emphasis on critical thinking skills can help them solve problems. In a real-world setting, students cannot rely on employers to provide solutions to difficult situations. In fact, new workers often cannot rely on anyone but themselves to solve the kinds of technical communication problems that can arise. We believe that reasoning through a series of problems, not to mention often vague instructions and misleading information, allows students to experience that real-world context that so many teachers strive to bring into the classroom.

The discipline of technical communication is built on a firm foundation of practical application and workplace context. In fact, technical communication is unique because of its focus on audience analysis, distinct format guidelines, appropriate writing patterns, and level of technical expertise. Because of this, we have created a book highlighting the core of this important and still evolving discipline: (1) scenarios that emphasize critical thinking skills and workplace realities and (2) fundamental rhetorical considerations necessary for successful response to the scenarios. Thus, this book is different from conventional technical communication texts in that all rhetorical guidelines have been condensed into the basics necessary to deal with problems. Students can, as a result, refer to the introduction of each chapter for instructions, format considerations, general rhetorical standards, and broad recommendations for document design. The rest of each chapter is devoted to the scenarios themselves.

We offer in this book original scenarios that instructors can use individually. The long soil incineration scenario that runs through every chapter allows instructors to teach memo and letter writing, technical definition, description, instruction writing, and report and proposal writing. Should instructors wish to use a different kind of scenario for each chapter, the book is arranged so instructors can choose from four other unique and thought-provoking situations for each new writing assignment.

We believe that, given the rapid advances in technology, the popularity of the Internet and electronic mail, and the constantly changing workplace environment, this approach—using scenarios—has never been more relevant in the technical communication classroom. Though the scenarios themselves will not solve the problems students are likely to face after graduation, such fictional situations do allow instructors to work with students on the aspects of writing in the workplace—collaboration, reasoning, critical thinking, and, perhaps most importantly, writing with a strong sense of audience and purpose.

ACKNOWLEDGMENTS

Though colleagues certainly prepared us for the rigors involved in creating a textbook, we could not have foreseen (a) the labor necessary to produce such a book and (b) the endless generosity of time and spirit certain individuals provided throughout the process.

Teresa Kynell thanks the many graduate students who assisted her in the process of developing the following chapters, especially Gina Derner (for her fantastic computer skills and visuals), Marcia Parkkonen, David Gardner, and all those who offered encouragement and office help along the way. Special thanks to Rose Rosina, one special secretary. And much gratitude to my colleague Anne Youngs who cheered me throughout the process and to my department head, Leonard Heldreth, who supported me when I needed extra time to complete this project.

For their assistance and constructive criticism early in the process, thanks must go to Katherine Staples and John Brockmann, two colleagues whose opinion is valued. Other instructors who provided valuable advice during the development of the manuscript are: Carol Barnum, Southern Polytechnic State University; Ray Dumont, University of Massachusetts, Dartmouth; Dan Jones, University of Central Florida; and Nancy L. Stegall, DeVry Institute of Technology. Certainly this business of acknowledging would not be complete without the mention of two wonderful friends, Fiona Gibbons and Donna Silta, who were there in good times and bad. And to Max and Olivia, a constant source of inspiration. Thanks also to Joe Opiela and Mary Beth Chesbrough both of Allyn and Bacon.

Finally, Teresa Kynell wishes to acknowledge with great gratitude the contribution of her best friend, Dr. Kurt V. S. Kynell. His support, superb copyediting, technical assistance, and visuals were of incalculable value. And now, *Patrician*.

Wendy Stone gratefully acknowledges and thanks the students of her Technical and Report Writing classes who worked through many of these scenarios and provided invaluable feedback. Thanks also to the colleagues and friends who provided insight, help, and ideas. Thanks go particularly to Daniel Stone, Melissa Borden, Robert Richardson, Sean Hayes, Robert Glenn, Raymond Ventre, Lillian Heldreth, Emily Godec, Amber Storey, Christine Swadley, and the entire Kabe Migeezeeug community. Special thanks to Mike Schweigert for the technical and other support. Last but not least, the greatest appreciation to Dr. Teresa Kynell for her years of tireless support and teaching, for her assistance during my thesis research, and for the honor of working with her.

T.C.K.
W.K.S.

CHAPTER 1

Using Scenarios to Learn Technical Communication

WHAT ARE SCENARIOS?

Scenarios, as you will study them, involve making decisions without knowing the outcome—that is, working on problems happening "now" and not knowing if your solution will work. You will also discuss problems with your "colleagues" (other students), draft documents, revise decisions, and defend what you choose to do. The emphasis in this textbook is not on second-guessing what happened to other people in other situations but on making your own decisions (often tough ones) *as if you were really in that particular situation.*

The scenarios in this text are designed to teach the most important aspect of technical writing—how to reach your audience. In other writing courses, you may have created personal stories or essays about thoughts and feelings on specific issues. Perhaps you evaluated the literature of others, finding connections between life as you know it and that presented by the author. Technical writing, though, is primarily concerned with what readers think and feel. Good technical writers navigate the murky waters of personalities, office politics, and ethics to produce clear, concise, effective communications that cause readers to take action in a specific way—the action you, the writer, desire them to take. For example, a technical writer preparing safety manuals for automobile assembly line workers needs to be very clear about what workers should and should not do because mistakes could result in loss of a limb or life if the technical writer does not consider all the issues. Beyond the technical aspects of the assembly line (the machinery involved), the technical writer must take into consideration the workers' different educational backgrounds, work experiences, and attitudes. This emphasis on readers is *audience analysis.* The scenarios in this text contain dialogue and other important information to help you *understand* readers so that your communication strategies are effective.

WHAT WILL SCENARIOS HELP YOU LEARN?

When you signed up for this course, you probably thought that it involved writing reports, creating a resume, and perhaps spending some time in the library conducting research. While these things certainly will occur, you will be learning much more through the scenarios in each chapter. Scenarios are designed to teach you many different skills—how to write a certain kind of document; evaluate ethical dilemmas; sort through erroneous, conflicting, or vague information; analyze readers' needs; and predict the results of your actions. Because scenarios may be new to you, we will take a look at some techniques for working through each scenario.

Tips for Understanding Scenarios

When reading a scenario, remember that every detail is important. Each scenario begins with a series of questions to help you focus your attention on specific pieces of information. In addition to keeping these questions in mind as your read, be sure to go over each scenario at least twice while doing the following:

- *Take Notes.* What information did you learn from the scenario? What *kind* of information did you receive (solid, vague, conflicting, suspect)? Who told you what? What are all the possible reasons your sources may have for providing you with the information? For example, a manager, tells you his division does not have enough money to launch a project you are proposing. You later find out that his division has extra money. Why, do you think, he said he had no funds?
- *"Listen."* How did the people in the scenario *seem?* What were their emotions and attitudes? Did someone's demeanor change, and when did that happen? For example, when you start asking questions about an accident at your company, people who were previously very friendly and helpful become cold and won't give you straight answers. Does this possibly reveal anything about the information they are providing?
- *Sketch the Relationships.* Create a drawing showing the relationship structure of the organization in the scenario. To whom do people report? Who has the most authority? The least? Who works together? Who has the power to do what? Which people are friends? Which are in conflict with each other? Sometimes you need to make sure you are not creating new problems, such as recommending that two people who dislike each other work together on a project. Perhaps your recommendation won't be accepted just on the basis of this personality clash.
- *List the Possibilities.* Decide what kind of document(s) you plan to write and then focus on content. What will you say? How will you

say it? How will you organize your information? In what other ways could you write this document? For example, you may have decided you need to write a letter to several employees informing them that their positions have been eliminated. Are you going to "let them down easy"? Will you simply tell them in a "just the facts" manner? Could you say anything that would help them understand the reason for this action? Will you offer assistance, such as a reference or a letter of recommendation?

- *List the Possible Outcomes.* For each possibility you have considered, try to forecast what will happen. How will readers feel about what you've written? What actions do you expect them to take? Could they possibly take other actions that would be counterproductive? For example, when negotiating a contract to provide computer training for a company, you find out that company is in the process of hiring a full-time staff member to train employees in exactly the same subject. If they go through with the hiring, there's really no need for your services, although you think the job is too big for just one person. Can you convince the company to sign a contract with your training group instead of hiring their own trainer?
- *Examine the Ethical Issues.* Did you receive information you promised not to repeat? Are you being asked to do something that may be, if not illegal, morally wrong? (And what do you consider morally wrong?) Would any of your actions result in harm to anyone, professionally or personally? For example, a friend of yours works in the personnel office, and he told you that your boss was recently reprimanded. Your friend tells you that if your boss makes any more mistakes, she will be fired and you will receive a promotion to her position. Just that morning, your boss misplaced some important documents and could not find them. Do you tell your friend about this mistake in hopes of getting the promotion?

HOW DO YOU FIND THE "RIGHT" ANSWER TO A SCENARIO?

Making the Best Possible Choices

The most important skills you use while working through scenarios are critical thinking and logical argumentation. You must have reasons for your decisions, and you must be able to articulate these reasons. "Just because I think so" is not a good argument. Use the information in the scenario and your own knowledge of technical communication to make your decisions. You may not "add" to the scenario, such as making up conversations, events, or facts that do not actually appear in the text.

The most important point to keep in mind is that you must decide what actions to take if you were in the situation described. As a result, expect your solution to be different than your classmates'. Because everyone's decision-making skills are built from years of making choices in many different situations, you have a unique way of looking at each scenario. Based on your personality, experiences, and decisions made during discussions in class and with your peers, you will take certain actions. Do you believe in being conservative in difficult situations? Are you a risk-taker? Are you willing to sacrifice people for profits? Are you more concerned with job satisfaction than position or pay? Your answers to these questions will shape your solutions to each scenario.

You will, therefore, write each document from the vantage point of *yourself in the organization described in the scenario, not as a student in a technical writing class.* For example, if the scenario says, "You are a technical writer for Montcalm Insurance," you would write your documents as "[Your Name], Technical Writer, Montcalm Insurance." Remember that your instructor will be reading the scenario *as the person or people to whom you are writing.* Your instructor will evaluate your response by examining how well you recognized and addressed the various issues in the scenario and will help you learn better decision-making skills with each new scenario. While no one "right" answer exists, understand that this does not mean that all scenario responses are valid—some will be grounded in a stronger critical framework than others. Keep in mind that grammar, mechanics, usage, and format are just as important as your message and *must* be correct—so plan to revise and proofread carefully.

Defending Your Answer to a Scenario

Once you have read a few scenarios and drafted documents in response to them, you are ready to address a second, very important issue. You must be able to explain *why* you made the decisions that resulted in your document or documents. Your explanation, called a **Solution Defense,** determines just how well your answer solves the scenario's problems. A Solution Defense demonstrates your ability to identify the scenario's problems and use the appropriate technical communication skills to solve them. Also, it illustrates your thinking processes—how well you took the information in the scenario and sorted it out. Essentially, a Solution Defense is an argument for *why your answer is the best one.*

Figure 1.1 lists all the typical aspects of a Solution Defense. While providing this basic information for each scenario, you should modify the format as you see fit by adding more information. Remember to write a Solution Defense for each assigned scenario.

AUDIENCE	• Who requested the document(s)? • Who "signs off" on it? • Who will read it? • Who may read it for another purpose? • In my organization, what positions do my readers hold? • What relationship do I have to my readers? • What information will my readers understand easily? • What will I have to explain? • In how much detail will I have to explain complex information? • What attitude do the readers have toward the document(s)? • What objections might they have toward what I'm asking? • Why would they have these objections? • How do I address those objections in my document(s)?
DOCUMENT TYPE	• What kind of documents (letter, memo, report, etc.) should I write for my readers? • Why use that kind of document?
DOCUMENT PURPOSE	• What, specifically, do I want my readers to do after they receive these documents? • When do I want them to take action? • What plan for taking action have I provided? • For what other purposes could my document be used?
DOCUMENT ORGANIZATION	• How did I organize my information? • Why did I choose to organize it that way?
SOLUTION EXPLANATION	• What, briefly, did I decide to do to solve the scenario's problems? • What are my reasons behind this solution? • What other actions did I consider but reject? • Why did I reject those possibilities? • What ethical problems did I encounter? • How did I work through them?
OUTCOME FORECAST	• What are the most possible reactions to my documents? • What would I do to address these possible reactions if given a chance to write another document in response?

FIGURE 1.1 Solution Defense Content

WORKING THROUGH A SCENARIO: A SAMPLE SITUATION

While using scenarios may seem difficult at first, it becomes easier as you become accustomed to the style. Read "The Disappearing Lake" scenario and take notes on your own. Then read the scenario as it was annotated by Amy (a technical writing student), read her notes, and take a look at her solution defense and documents. While Amy's solution may not be what you would have done, does she provide enough information for you to understand her reasoning?

The Disappearing Lake

CONSIDERATIONS

In this situation, you are a reporter working in a small town near a recently closed Army base. You hear rumors about toxic waste and cleanups and about a possible connection between activities on the base and the water levels of two lakes. As you read this scenario, consider the following:

- What are the possible reasons for Deer Lake's loss of water?
- What type of evidence is provided for a connection between activities on the Army base and a drop in Deer Lake's water level?
- How does your status as a home owner and reporter affect the credibility of any documents you produce?

You pull up in front of your new home on the shores of Deer Lake, thankful that the last load from the old house seems to have survived, undamaged, the ride in the back of your truck. You grab the heaviest box first, hoping to get the worst over with before the August morning turns into a scorching afternoon, and head for the garage.

"Hello there!" a voice booms, startling you enough to loosen your grip on the box as it crashes to the ground. Books spill out from the broken cardboard. To your right, a man is standing on the other side of your neighbor's low wooden fence. Noting his slight build, gray hair and red T-shirt, you wonder if he is the elusive neighbor you haven't seen during the past week of moving.

"Hello," you answer, rather flatly, a little frustrated with the mess you've made. The man easily hops the fence, walks across the small patch of browning grass between the fence and your driveway, and begins helping you collect the spilled books.

"Sorry I scared you," he says. "I just wanted to get acquainted with my new neighbor. I'm Don Brayburn, president of the Deer Lake Homeowner's Association."

"Nice to meet you, Don," you respond, and you give your name. Turning to pick up the books, you continue, "I probably would have dropped them anyway. Someday, I should give away all these books. I don't have time to read them anymore."

"Hhmmm . . . " Don eyes the covers of the books he's picked up as the two of you walk toward the garage. "Emerson, Whitman, and Dickens. Good stuff. Well, I heard you were a journalist," Don says, "so I wondered just what kind you were. Too many liberal journalists around these days."

You realize that if you say what you're thinking, a neighborhood feud might commence. You decide to keep your mouth shut and be as accommodating as possible. "I'm a newspaper journalist who likes hard facts and small towns," you say. "I don't think you have anything to worry about from me."

"Good," Don says. "There's enough trouble around here, what with the lake disappearing."

"Deer Lake's disappearing?" you ask, surprised. The realtor who sold you the house said nothing about this.

"It's gone down about a foot since last year," Don says. "Ever since they've been dredging a lake at the Army base, all the lakes around here have been shrinking."

Because this might be a potential story for the *Daily Mirror,* the local newspaper for which you work, you invite Don into the house for a glass of iced tea. He accepts and soon you are sitting on your backyard deck, enjoying the cool breeze blowing off the lake. For the first time, you notice several inches of black mud between your lawn and the water's edge. You wonder if it's an indication of how much the water has dropped. Eventually, you find a way to get Don to talk about the lake again after he'd gone on for twenty minutes about his exploits in the military.

"A buddy of mine, a captain, told me that the Army dumped into the small base lake, Lake McCarthy, for years, and now that the base is closing, they're trying to clean it up," Don says. "He said they drained the lake and are digging up the muck at the bottom. There's guys in orange suits working the machinery, and they're putting the muck in big plastic drums, trucking it out only at night and under tarps, at that. The Army started draining in late May, and I noticed the water level drop in early June."

"What's your buddy's name?" you ask, thinking that this guy could be a good source.

"Sorry," Don says, smiling. "I told him I'd keep it hush-hush. You want to do a story on this?"

"Sure," you say. "It sounds like there could be something to this. But I need more information."

"I can give you information, contacts, whatever," Don says. "But I've talked to the *Mirror's* editor twice, and she says the paper's not interested."

You wonder why your editor, Linda Mann, declined to pursue such a potentially big story. There must be something here more than you know. . . .

Don interrupts your train of thought. "Call Colonel Firestein at the base," he says. "Tell him you know me, and maybe he'll give you an official answer to questions about the dredging. I haven't asked him. We're good friends, and I don't want to put him in a funny spot."

You nod, remembering the countless times you'd refrained from asking friends for information, even though you know they had what it would take to make a good story. Small-town journalism is sometimes very difficult. "I'll call him," you assure Don as he gets up to leave.

Later that week, you've worked through the pile of paperwork on your brand new desk at the *Daily Mirror* and find time to tackle the Deer Lake story. You call the information number at Connelly Army Base and are connected to Colonel Firestein's office. After mentioning Brayburn to the secretary, she puts you through to the Colonel.

"Yes!" a loud voice booms into your ear, and you hold the phone further away.

"Colonel Firestein, I'm calling from the *Daily Mirror,*" you say. "I have some questions about environmental cleanups going on at the base."

"That's not really my area," the Colonel says. "But since Don told you to call me, I'll try to answer your questions."

"Have you drained Lake McCarthy?" you ask.

"Yes," he answers, and offers nothing more.

"Why?" you ask, matching his monosyllabic answer with a similar question.

"Back in the forties and fifties, we dumped used oil in there. It wasn't really a lake to begin with, more like a swamp the Army dug out to use as a holding area for waste fluids. There weren't regulations on how to do that kind of thing back then. The EPA knows all about this."

"What are you doing with the material you dredge from the lake bottom?" you ask.

"Who says we're dredging?" the Colonel shoots back.

"Well," you respond, "it's only logical, since you drained the lake. And the fact that the lake sediment would be toxic if you dumped waste fuel into it for the last fifty years."

"Humph," the Colonel grunts. "We're trucking the stuff to a soil incineration plant."

"At night?" you ask, thinking that the Colonel knows quite a bit about these issues, despite it not "being his area."

"Regulations," he answers, without questioning how you knew this small fact. "It's safer without all the traffic."

"I understand," you say, preparing to move on to touchier ground. "Did the Army conduct any studies on how draining Lake McCarthy would affect the local water table?"

"I believe so, but don't quote me on that," the Colonel answers. "Why do you want to know?"

"Because Brayburn claims Deer Lake's lost about a foot of water, and that it happened at the same time the Army drained Lake McCarthy."

"Nonsense," the Colonel said. "The lakes around here go up and down all the time, have for years. We're in the middle of a dry summer."

"Could I obtain a copy of the environmental study done before the cleanup began?" you ask.

"Look, I have a meeting to get to. Ask my secretary for the report, and we'll see if it's available to the public. Good-bye."

The Colonel hung up without transferring you to his secretary, so you had to call back and make your request. From her tone of voice, you doubt if you'll ever see the report. You hang up the phone and lean back in your chair, staring out the window at the busy main street below.

"Hey, there!" Linda's voice startles you. "What, not enough work to do?"

"I was thinking," you said, smiling. You like Linda—she's a good, ethical editor concerned more with getting the story right than with circulation numbers.

"About?" she asks. You laugh, wondering how you'll keep any secrets in an office full of reporters.

"I just got off the phone with Colonel Firestein at the Army base," you answer, and then you fill her in on what you've learned so far. You leave out what Don told you about Linda declining to pursue the story.

"You got farther than I did," Linda said. "I think we should run a story on the base's cleanup efforts. It would be good to let the community know everything will be okay out there after the Army leaves. Why don't you call around and write something up on it?"

"Sure," you say. "But what about the lake level issue? Isn't that pertinent to the property owners?"

"Yourself included," she reminds you. "Conflict of interest. Stay away from that angle. Besides, Deer Lake hasn't lost any more water than it normally does on a dry year. And Brayburn's always talking conspiracy."

"Really," you say, sounding not at all surprised. You grab your notepad and pencil.

"Where are you going?" Linda asks.

"Library," you reply. "I'm going to do some research on the deer population story."

"Have fun," Linda says as you leave.

After tracking down deer herd population statistics, you decide to stay at the library for a few more minutes and look up the records on Deer Lake. You find several studies done by the Department of Natural Resources on the lake, mainly dealing with its algae-bloom problem. You find an algae-bloom report that summarizes lake level data for many years as part of other calculations, so you take down some notes:

Depths are averages of measurements taken in the general vicinity of the lake middle where currents from supply streams should not greatly affect lake bottom contours. Measurements conducted first week in July. Only have data from 1978 on.

1978—75 ft	88—70 ft
79—76 ft	89—72 ft
80—76 ft	90—72 ft
81—75 ft	91—73 ft
82—75 ft	92—73 ft
83—74 ft	93—72 ft
84—72 ft	94—72 ft
85—71 ft	95—73 ft
86—70 ft	96—74 ft
87—72 ft	97—74 ft

Satisfied that the lake will more than likely rise again once the summer is over, you return to the office. Just as you sit down at your desk, your phone rings. It's Don.

"Firestein just called me," he says. "He was all mad about you asking him so many questions and mad at me for giving you his name."

"Really?" you say. "Did he say anything else?"

"Just that I'm wrong about the lake. But I'm going to force the issue."

"How?" you ask.

"Last night, at the Deer Lake Home Owner's meeting, I announced that you were investigating our problem and that you'd be running a story in the *Mirror* on it!"

"Oh no," you groan. "Don, I can't do the story. I'm automatically a member of the association because I own property on the lake. And what's more, I haven't found anything to suggest the lake is any lower than it normally gets on a dry year."

"Too late," Don says. "And who better to write the story than someone whose home is at stake?"

"You don't understand," you try to explain. "No paper would run a story written by someone with such an obvious bias. Tell them I can't do it."

"Then you tell them you'll get us someone who will," Don answers with a sharpness in his voice. He also hangs up the phone.

You realize that your reputation in this town is on the line. *I could lose people's trust in my professionalism,* you think. *And the first thing people would hear about the new reporter in town would be this controversy.*

You recall a lesson your favorite journalism professor pounded into your head during your college years: "People remember the first things they hear." You are very upset at Don, and you decide you must put something in writing, go on record, to make your position clear to all involved.

Now evaluate Amy's response to the scenario shown in Figure 1.2. First, look at how she made notes, or annotations, to the case as she read it. "Marking up" a case is important, as it helps you identify key pieces of information that you will need to reference in your Solution Defense.

The Disappearing Lake

CONSIDERATIONS

In this situation, you are a reporter working in a small town near a recently closed Army base. You hear rumors about toxic waste and cleanups and about a possible connection between activities on the base and the water levels of two lakes. As you read this scenario, consider the following:

- What are the possible reasons for Deer Lake's loss of water?
- What type of evidence is provided for a connection between activities on the Army base and a drop in Deer Lake's water level?
- How does your status as a home owner and reporter affect the credibility of any documents you produce?

Conflict of interest?

You pull up in front of your new home on the shores of Deer Lake, thankful that the last load from the old house seems to have survived, undamaged, the ride in the back of your truck. You grab the heaviest box first, hoping to get the worst over with before the August morning turns into a scorching afternoon, and head for the garage.

"Hello there!" a voice booms, startling you enough to loosen your grip on the box as it crashes to the ground. Books spill out from the broken cardboard. To your right, a man is standing on the other side of your neighbor's low wooden fence. Noting his slight build, gray hair and red T-shirt, you wonder if he is the elusive neighbor you haven't seen during the past week of moving.

FIGURE 1.2 Amy's Annotations to the Scenario

"Hello," you answer, rather flatly, a little frustrated with the mess you've made. The man easily hops the fence, walks across the small patch of browning grass between the fence and your driveway, and begins helping you collect the spilled books.

dry weather

"Sorry I scared you," he says. "I just wanted to get acquainted with my new neighbor. I'm Don Brayburn, president of the Deer Lake Homeowner's Association."

gives his title?

"Nice to meet you, Don," you respond, and you give your name. Turning to pick up the books, you continue, "I probably would have dropped them anyway. Someday, I should give away all these books. I don't have time to read them anymore."

"Hhmmm . . ." Don eyes the covers of the books he's picked up as the two of you walk toward the garage. "Emerson, Whitman, and Dickens. Good stuff. Well, I heard you were a journalist," Don says, "so I wondered just what kind you were. Too many liberal journalists around these days."

Don is a conserva-tive?

You realize that if you say what you're thinking, a neighborhood feud might commence. You decide to keep your mouth shut and be as accommodating as possible. "I'm a newspaper journalist who likes hard facts and small towns," you say. "I don't think you have anything to worry about from me."

"Good," Don says. "There's enough trouble around here, what with the lake disappearing."

"Deer Lake's disappearing?" you ask, surprised. The realtor who sold you the house said nothing about this.

"It's gone down about a foot since last year," Don says. "Ever since they've been dredging a lake at the Army base, all the lakes around here have been shrinking."

Lake level drops

Because this might be a potential story for the *Daily Mirror,* the local newspaper for which you work, you invite Don into the house for a glass of iced tea. He accepts and soon you are sitting on your back-yard deck, enjoying the cool breeze blowing off the lake. For the first time, you notice several inches of black mud between your lawn and the water's edge. You wonder if it's an indication of how much the water has dropped. Eventually, you find a way to get Don to talk about the lake again after he'd gone on for twenty minutes about his exploits in the military.

No real source—unreliable evident

"A buddy of mine, a captain, told me that the Army dumped into the small base lake, Lake McCarthy, for years, and now that the base is closing, they're trying to clean it up," Don says. "He said they drained

FIGURE 1.2 Amy's Annotations to the Scenario *continued*

toxic waste procedures? the lake and are digging up the muck at the bottom. There's guys in orange suits working the machinery, and they're putting the muck in big plastic drums, trucking it out only at night and under tarps, at that. The Army started draining in late May, and I noticed the water level drop in early June."

"What's your buddy's name?" you ask, thinking that this guy could be a good source.

"Sorry," Don says, smiling. "I told him I'd keep it hush-hush. You want to do a story on this?"

"Sure," you say. "It sounds like there could be something to this. But I need more information."

that I can use? "I can give you information, contacts, whatever," Don says. "But I've talked to the *Mirror's* editor twice, and she says the paper's not interested."

You wonder why your editor, Linda Mann, declined to pursue such a potentially big story. There must be something here more than you know. . . .

But doesn't mind you (me) doing it? Don interrupts your train of thought. "Call Colonel Firestein at the base," he says. "Tell him you know me, and maybe he'll give you an official answer to questions about the dredging. I haven't asked him. We're good friends, and I don't want to put him in a funny spot."

You nod, remembering the countless times you'd refrained from asking friends for information, even though you know they had what it would take to make a good story. Small-town journalism is sometimes very difficult. "I'll call him," you assure Don as he gets up to leave.

Later that week, you've worked through the pile of paperwork on your brand new desk at the *Daily Mirror* and find time to tackle the Deer Lake story. You call the information number at Connelly Army Base and are connected to Colonel Firestein's office. After mentioning Brayburn to the secretary, she puts you through to the Colonel.

"Yes!" a loud voice booms into your ear, and you hold the phone further away.

"Colonel Firestein, I'm calling from the *Daily Mirror,*" you say. "I have some questions about environmental cleanups going on at the base."

"That's not really my area," the Colonel says. "But since Don told you to call me, I'll try to answer your questions."

"Have you drained Lake McCarthy?" you ask.

"Yes," he answers, and offers nothing more.

"Why?" you ask, matching his monosyllabic answer with a similar question.

FIGURE 1.2 Amy's Annotations to the Scenario *continued*

"Back in the forties and fifties, we dumped used oil in there. It wasn't really a lake to begin with, more like a swamp the Army dug out to use as a holding area for waste fluids. There weren't regulations on how to do that kind of thing back then. The EPA knows all about this."

—could get EPA report?

"What are you doing with the material you dredge from the lake bottom?" you ask.

"Who says we're dredging?" the Colonel shoots back.

"Well," you respond, "it's only logical, since you drained the lake. And the fact that the lake sediment would be toxic if you dumped waste fuel into it for the last fifty years."

"Humph," the Colonel grunts. "We're trucking the stuff to a soil incineration plant."

"At night?" you ask, thinking that the Colonel knows quite a bit about these issues, despite it not "being his area."

Believable.

"Regulations," he answers, without questioning how you knew this small fact. "It's safer without all the traffic."

"I understand," you say, preparing to move on to touchier ground. "Did the Army conduct any studies on how draining Lake McCarthy would affect the local water table?"

"I believe so, but don't quote me on that," the Colonel answers. "Why do you want to know?"

"Because Brayburn claims Deer Lake's lost about a foot of water, and that it happened at the same time the Army drained Lake McCarthy."

"Nonsense," the Colonel said. "The lakes around here go up and down all the time, have for years. We're in the middle of a dry summer."

No documentation available from the Col.

"Could I obtain a copy of the environmental study done before the cleanup began?" you ask.

"Look, I have a meeting to get to. Ask my secretary for the report, and we'll see if it's available to the public. Good-bye."

The Colonel hung up without transferring you to his secretary, so you had to call back and make your request. From her tone of voice, you doubt if you'll ever see the report. You hang up the phone and lean back in your chair, staring out the window at the busy main street below.

"Hey, there!" Linda's voice startles you. "What, not enough work to do?"

"I was thinking," you said, smiling. You like Linda—she's a good, ethical editor concerned more with getting the story right than with circulation numbers.

"About?" she asks. You laugh, wondering how you'll keep any secrets in an office full of reporters.

FIGURE 1.2 Amy's Annotations to the Scenario *continued*

"I just got off the phone with Colonel Firestein at the Army base," you answer, and then you fill her in on what you've learned so far. You leave out what Don told you about Linda declining to pursue the story.

"You got farther than I did," Linda said. "I think we should run a story on the base's cleanup efforts. It would be good to let the community know everything will be okay out there after the Army leaves. Why don't you call around and write something up on it?"

"Sure," you say. "But what about the lake level issue? Isn't that pertinent to the property owners?"

No story

"Yourself included," she reminds you. "Conflict of interest." Stay away from that angle. Besides, Deer Lake hasn't lost any more water than it normally does on a dry year. And Brayburn's always talking conspiracy."

Don's reliability is in question.

"Really," you say, sounding not at all surprised. You grab your notepad and pencil.

"Where are you going?" Linda asks.

"Library," you reply. "I'm going to do some research on the deer population story."

"Have fun," Linda says as you leave.

After tracking down deer herd population statistics, you decide to stay at the library for a few more minutes and look up the records on Deer Lake. You find several studies done by the Department of Natural Resources on the lake, mainly dealing with its algae-bloom problem. You find an algae-bloom report that summarizes lake level data for many years as part of other calculations, so you take down some notes:

focus of study was not lake level.

Depths are averages of measurements taken in the general vicinity of the lake middle where currents from supply streams should not greatly affect lake bottom contours. Measurements conducted first week in July. Only have data from 1978 on.

No lake level problem. it seems...

1978—75 ft
79—76 ft
80—76 ft
81—75 ft
82—75 ft
83—74 ft
84—72 ft
85—71 ft
86—70 ft
87—72 ft

88—70 ft
89—72 ft
90—72 ft
91—73 ft
92—73 ft
93—72 ft
94—72 ft
95—73 ft
96—74 ft
97—74 ft

FIGURE 1.2 Amy's Annotations to the Scenario *continued*

Satisfied that the lake will more than likely rise again once the summer is over, you return to the office. Just as you sit down at your desk, your phone rings. It's Don.

"Firestein just called me," he says. "He was all mad about you asking him so many questions and mad at me for giving you his name."

Trying to make himself credible through my "interest"?

"Really?" you say. "Did he say anything else?"

"Just that I'm wrong about the lake. But I'm going to force the issue."

"How?" you ask.

"Last night, at the Deer Lake Home Owner's meeting, I announced that you were investigating our problem and that you'd be running a story in the *Mirror* on it!"

"Oh no," you groan. "Don, I can't do the story. I'm automatically a member of the association because I own property on the lake. And what's more, I haven't found anything to suggest the lake is any lower than it normally gets on a dry year."

"Too late," Don says. "And who better to write the story than someone whose home is at stake?"

"You don't understand," you try to explain. "No paper would run a story written by someone with such an obvious bias. Tell them I can't do it."

"Then you tell them you'll get us someone who will," Don answers with a sharpness in his voice. He also hangs up the phone.

You realize that your reputation in this town is on the line. *I could lose people's trust in my professionalism,* you think. *And the first thing people would hear about the new reporter in town would be this controversy.*

You recall a lesson your favorite journalism professor pounded into your head during your college years: "People remember the first things they hear." You are very upset at Don, and you decide you must put something in writing, go on record, to make your position clear to all involved.

Don = unreliable
Col. Firestein = no documentation available
Linda = discourages me writing on this
me = found no real evidence of a problem
but am now in a somewhat controversy

FIGURE 1.2 Amy's Annotations to the Scenario *continued*

Examine Amy's solution defense shown in Figure 1.3. Notice how she identifies her readers and indicates what she believes are their motivations and needs concerning the Deer Lake situation. She then discusses the documents she created, provides reasons for her decisions, and evaluates the possible outcomes of her actions.

The Disappearing Lake Scenario Solution Defense

Amy Smith

Readers

My readers are the Deer Lake Homeowner's Association (DLHA) members, including Don Brayburn. Another reader is my editor, Linda Mann. Don will not want to hear what I have to say and neither will the other members of the DLHA. Linda needs to read this letter so she will know I'm not going after the story, contrary to popular rumor. She must be able to defend the paper against anyone saying we're being unethical by having me do a story about a lake on which I live. Linda won't be happy about the situation. It's possible that my letter could be read by people at the Army base (including Colonel Firestein) if someone from the DLHA gives it to them.

Document

I'm writing a letter to the DLHA and copying Linda on the letter. I'll write a short cover memo for Linda and attach the letter to it. I chose the letter format because I'm writing an official communication from the paper to the DLHA, so I'll use *Daily Mirror* letterhead. I'm writing a memo to Linda because of a purely internal purpose for that communication—to let her know what Don did and what I did to correct his mistake.

Solution

For the letter, I stated up front that Don was mistaken in reporting that I would be investigating and writing on the Deer Lake water level matter. I then gave my reasons in order of importance: the conflict of interest, the data that doesn't support his contention that water levels

FIGURE 1.3 Amy's Solution Defense

are dropping more than normal, and then the lack of hard evidence. I then referred them to Linda if they had any further things to discuss about the issue. For the memo to Linda, I started by reminding her about my investigation into the Deer Lake water level (because we had only talked about it briefly). I told her what Don did (problem), and then I mention the letter and my intention to direct all questions about Deer Lake to her (solution). I ask her to handle the situation and apologize for any negative effect the whole situation could have on the paper's reputation. I'm new, and I don't want to be known as a troublemaker right off.

Considerations

I want the DLHA members to not tell anyone that I am investigating the water level issue. I want Linda to be very happy with how I handled the situation so that there's no appearance of conflict of interest. I want Don to read the letter and get off my back about this issue, but I don't want him to get angry with me, because I have to live next to him for a very long time. I really don't care if Army personnel read my letter to Don, because there's no evidence that the base lake and Deer Lake are physically connected.

Outcomes

As for outcomes, I believe the letter will have varied effects on the DLHA members, depending on how prone to suspicion they are. I think most will understand why I can't pursue the story; however, I do believe some will think I'm now part of the cover up. I believe that's what Don might think. I'm hoping that my offering up Linda as a contact for this issue will fulfill Don's request for an investigator. Perhaps she can convince him that the lake level is fluctuating normally. I'm hoping that Don or other DLHA members tackle the items I listed as unknowns. Ultimately, I just want to save my reputation, my job, and not make my neighbors unhappy with me in the process. I am not concerned about any environmental problems with Deer Lake.

FIGURE 1.3 Amy's Solution Defense *continued*

Now that you've read her solution defense, read Amy's documents (Figs. 1.4 and 1.5) and determine whether or not her writing accomplished the goals she defined in her defense.

The Daily Mirror
202 Front Street
Anytown, USA 55555

August 10, 1999

Deer Lake Homeowner's Association
c/o Don Brayburn, Chair
204 Deer Lake Road
Anytown, USA 55555

Dear fellow members:

Although I am a very new member of the DLHA, I have recently become involved in a complex situation regarding the alleged reduction in water level in Deer Lake. Chair Don Brayburn asked me to investigate this matter as a possible story for the *Daily Mirror*, for which I am an employee (writer).

I began investigating this matter and spoke to Colonel Firestein at the Army base, and while he confirmed the Army is cleaning up waste in a small lake, he denies any connection between this activity and possible reduction in Deer Lake's water level. According to the Colonel, the Army conducted a geological study of the area's water table before commencing with the cleanup; however, I have not received a copy of this study. The Department of Natural Resources has been measuring the lake level since the late 1970s; after reviewing their data, I believe Deer Lake's water level is fluctuating normally.

As a writer for the paper, I must be unbiased, and my personal interests in Deer Lake, as a property owner, could be perceived as a conflict of interest. You may contact Linda Mann, my editor, with any new information. At this time, the *Daily Mirror's* position is that there is no story here.

I hope you understand the difficult position I am in as both a professional and a concerned Deer Lake homeowner. While I cannot assist you with your investigation, I look forward to becoming an active member of the association.

Sincerely,

Amy Smith

FIGURE 1.4 Amy's Letter to the DLHA

MEMORANDUM

TO: Linda Mann, Editor

FROM: Amy Smith

RE: Deer Lake Water Level Investigation

As requested, I have dropped my investigation into the Deer Lake water level/Army base cleanup connection. Unfortunately, the chair of the DLHA told members I was investigating the issue and writing a story on it.

Attached is a copy of the letter I wrote to the DLHA members. It should effectively detach me, and thus the paper, from any perception of conflict of interest. I provided the members with your name as the contact person should they acquire any new information.

Linda, I am truly sorry that one of my first investigations could tarnish the ethical image of the paper. I apologize for not using enough foresight to anticipate this turn of events. If you wish to discuss the matter further, please contact me.

FIGURE 1.5 Amy's Memo to Her Editor

Exercises for Classroom Discussion

1. Write a memo to your instructor, evaluating the effectiveness of Amy's scenario solution documents. Discuss what you would have done differently and why.

2. Write your own solution defense to the Disappearing Lake Scenario.

3. Write your own documents for the scenario.

4. Assume that Linda is unhappy with how you handled this matter. She wants you to write to Colonel Firestein to make sure he understands the paper's position on this issue. Write that document.

5. You decide to write a personal letter to Don, thinking that might help your status as neighbors. Write that document.

Now that you have evaluated one person's response to a scenario, you need to begin this critical thinking process. In Chapter 2, you will learn about the most common types of workplace communication—memorandums and letters. After you've mastered the format and content requirements for these types of documents, you will put them to use in one or more scenarios.

Until you become accustomed to working through scenarios, refer back to this chapter for assistance on forming your Solution Defense.

CHAPTER 2

Memoranda and Letters

Once you are involved in the day-to-day business of writing and making decisions, there is perhaps no one form of communication on which you will rely more often than memoranda and letters. Both memos (the abbreviated form of "memoranda") and letters are usually the most personal and immediate means available for you to reach colleagues, customers, clients, competitors, or anyone else with whom you do business. Memos and letters, though, are very different in purpose, intent, format, features, and type.

MEMOS

We'll begin with memos because they are the most often used form of communication within your business. Although electronic mail (e-mail) is rapidly becoming a popular means of communicating "in-house" (within a business or company), memos remain a staple of the workplace.

Memos are, essentially, *internal* correspondence. Anyone involved in the business world uses memos to communicate with colleagues or supervisors in the same company or organization. Thus, memos are somewhat less formal than letters but are still a very important means of information exchange. Memo content is defined by the *subject line,* which means that the writer keeps information focused on a central topic. Memos are typically brief (one page) unless a *memo report* is necessary, and then a memo might run as long as two or three pages. Memos, though short and somewhat informal, still require the same attention to audience, purpose, and context as other technical documents. This section covers general guidelines for formatting a memorandum and standards for the memo report.

Memorandum Format

Before we begin covering the types of memos you might write in the workplace, evaluate the example of correct memo format in Figure 2.1.

Notice the following distinct features of the memorandum:

1. You may center the word "MEMORANDUM" (in capitalized letters) across the top of the first page of your memo. You may also left justify this line.
2. Double space and type the TO line (to the person for whom the memo is intended). Remember to include a colon (:) after TO and tab over twice before entering the person's name.
3. Double space and type the FROM line (your name and title). Be sure to initial (a kind of memo "signature") the FROM line. Remember to include a colon (:) after FROM and tab over twice before entering your name.
4. Double space and type the DATE line. The DATE line is important for future reference, so be sure to use the correct date. Remember to include a colon (:) after DATE and tab over twice before entering the date. You may spell out the month followed by the day and year (example: January 2, 1999), or use the numeric dating method (1-2-99).
5. Double space and type the SUBJECT line to indicate the purpose of the memo. Make the SUBJECT line brief but illustrative of the primary point of the memo. Remember to include a colon (:) after SUBJECT and tab twice before typing the subject line. NOTE: You may also use "RE" in place of "SUBJECT."

In addition, remember the following general guidelines for writing memos:

- Make sure you have a clear purpose or intent for writing your memo. Reflect that purpose or intent both in the subject line and in the opening paragraph of your memo.
- Use headings to break up large blocks of information and to help you change from one subject to another.
- Arrange your material in a general-to-specific pattern. Make your primary point early and follow, in subsequent paragraphs, with more specific detail to support your main idea. Do not confuse *general* with *vague.* You should still provide relevant information in the more general parts of your memo.

Headers

MEMORANDUM

TO: All Employees

FROM: Alice Jones, Personnel Supervisor AJ

DATE: February 15, 1998

SUBJECT: Updated Resumes Required by March 1

Indication of purpose

Many of you may be aware of the contract ONEX Incorporated recently won with the Department of Natural Resources. The proposed project, a complete survey of 200 acres of government-owned forest in Alaska leading to a proposal for a new national park site, will take a collaborative effort. Company President Harlan Riggs has asked me to collect from all of you updated resumes for our files.

Heading introduces background information

Updating Personnel Files

We are updating company files for several reasons. First, one of the conditions in awarding us this lucrative contract was an updated resume of all employees involved in the project. Second, we need to evaluate both your college-level work experience as well as your subsequent work-related expertise before we assign specific jobs within the project. Third, resumes have not been updated for three years; this seems like a good time to take care of this.

Heading introduces the request for action

Guidelines for Updating Personnel Files

If you do not have a copy of your old resume, please see me for a file copy. Once you have the old copy, please update the following areas:

Bulleted list separates specific requirements

- work-related expertise (jobs or special projects you have completed since the last update)
- education (coursework, certificates, or degree programs)
- awards (grants, certificates, or any other accolade)
- special skills (any new and unique skills you have acquired in the last five years)

In addition, please prepare your resume using WordPerfect 6.0. Submit to me a hard copy with disk no later than March 1, 1998.

Closing

We appreciate your cooperation on this project.

FIGURE 2.1 Correct Memorandum Format

- Refer to specific individuals by name when relevant and use the pronouns *you* and *your* in addressing the recipient.
- If you are presenting a list of items, break the list away from the text and either number or bullet separate items. Avoid long blocks of text, particularly when you have items that can be listed.
- If you are making a scenario for action, consider saving your strongest point for the end of the memo.
- Always write a sentence or two of summation to conclude your correspondence. Don't leave your reader dangling by simply ending your memo on a main point. This is a good time to direct you reader to the first step in taking the action you've requested.

TIP

Most importantly, remember that memos are rarely read carefully, particularly by people who receive many in one day. Thus, as a memo writer you must make clear, in both the subject line and the first paragraph, your main point. Memos should be an exercise in efficiency, clear expression, and carefully reasoned, persuasive writing.

Memo Reports

In addition to being an efficient method of internal communication, memos can also be used in *memo reports.* Memo reports may take on many forms, but generally all types are used to persuade someone in a position of power to consider your ideas. Thus, memo reports recommend, suggest change, report information, justify an activity or expenditure, indicate progress on a project, and request assistance.

All memo report types are slightly different, but all follow the general guidelines for memo writing. The only difference in the memo report is (a) the length of the memo and (b) the more thorough coverage of information in the memo.

Memo reports should run approximately two to three pages; you should write a formal report for documents longer than three pages (covered in Chapter 8). You arrange memo reports in a general-to-specific pattern, but include more detail in the body portion of the memo. In addition, memo reports usually contain more headings and information and generally conclude with a request for action. Consider the memo report in Figure 2.2, a request for funding to purchase advertising space.

MEMORANDUM

TO: Robert Hill, Graduate Dean

FROM: Barbara Bennett, English Graduate Director

DATE: March 18, 1999

RE: Advertising Campaign Funding

Indication of purpose

This report covers the expenditures associated with our proposed English MA advertising campaign. I have reviewed our previous plan, made suggestions for change, and proposed a new plan for the 1999–2000 academic year. In addition, this report addresses both the advertising strategies currently employed by the Department of English and those strategies recommended for future ad campaigns.

The Current English Advertising Plan

Background information

Examination of our current plan reveals two strategies: an $8^{1}/_{2} \times 11$ inch flier and brochure sent to other schools and, sometimes, a classified ad sent to the *American Poets and Writers* journal. The flier contains only information pertinent to the program: deadlines, stipends, and a contact name. The brochure, designed and printed in the late 80s, is efficient and contains relevant information. Somewhat more sophisticated than the flier, the brochure contains some color and formatting. The brochure, however, repeats the information on the flier.

Analysis of background

Our very small classified ads, printed alongside other such ads, attract very little attention. In fact, the percentage of responses based on such ads, according to research on advertising, is typically low. People, frankly, are attracted to the *appearance* of ads as well as the positive features expressed in the ad. Apparently the English Department has relied solely upon classified ads, in the past, to attract potential students.

FIGURE 2.2 Memorandum Report

Overview of request

A New Advertising Strategy

I propose that for the 1998 academic year, the English Department consider a moderate revamping of the advertising campaign. A moderate revamping will require a greater financial expenditure (though not significantly so). I believe we should use a larger poster format for mailings to other schools and a very different ad for two key journals.

Specific request information

New English Department Poster

I recommend that we eliminate the flier/brochure mailing and use instead an 11 × 17 inch poster advertising the program. We can develop the poster with little additional cost because students in the program are willing to design the poster.

The poster will focus on our region, emphasizing the natural beauty, our scholarship opportunities, our low student/professor ratio, and our special opportunities. The poster will include a photograph representative of our region (taken by a student photographer).

Journal Advertisement

Previously we limited our journal advertisements to classified ads in very small print with only basic information. I propose a more stylistic ad depicting the beauty of our region as well as the "hidden" benefits of an MA in English. Running such an ad will cost more money; producing the ad, though, is free because of student involvement. I propose that we run the ad in two journals: *American Poets and Writers* and *Writing Magazine.*

The subsection "Journal Advertisement" is further divided to provide detailed information on each proposed journal advertisement.

American Poets and Writers

This tabloid style journal, supported by Roosevelt University, is well known to writers. Its exact readership is unknown but is certainly around 30,000. The ad rates are as follows:

Full-page ad	$876
Half-page ad	$457
Quarter-page ad	$239

A quarter-page ad should be big enough for our purposes.

FIGURE 2.2 Memorandum Report *continued*

Writing Magazine

Another popular journal of creative writing, *Writing Magazine,* also runs display ads for MA programs. Boasting a readership of over 190,000, this journal is a little more expensive. The ad rates are as follows:

Full-page ad	$1570
Half-page ad	$820
Quarter-page ad	$450

Conclusion

Again, a quarter-page ad should work for our purposes.

Invitation for further discussion

In conclusion, I estimate that we should be able to advertise our program, with a new poster and three ads, for roughly $1000. I realize that this is more than we typically spend, but the experiment may prove beneficial with increased interest in our program. If you have any questions or would like to discuss this ad campaign in greater detail, please give me a call.

FIGURE 2.2 Memorandum Report *continued*

LETTERS

Letters, unlike memos, are the most often used form of communication outside your business. Letters are *external* correspondence and are thus more formal than memos. Letters, like memos, are typically brief and usually express one main point. Like memos, letters require the same attention to audience, purpose, and context as other technical documents. You may find, once you are on the job, that you will need to communicate certain *types* of information in your letters. This section will cover general guidelines for formatting a business letter and standards for different types of letters.

Letter Format

In industry and business today, we most often use two letter *styles:* block and modified block. Before we begin covering different types of letters you might write in the workplace, evaluate the two models (Figures 2.3 and 2.4 on pages 30 and 31) of correct letter format.

Date	January 12, 1997
Inside address State is fully spelled	Ms. Martha Andrews, Manager Data Services, Inc. 2300 Broadway Ave. Louisville, Kentucky 40272
Greeting	Dear Ms. Andrews:
Purpose of the document	Recently our new company, Ship With Us, began providing freight, shipping, and hauling services to the residents of western Oklahoma. While we believe our services are as prompt and efficient as that of anyone in the field, the first six months of business has us wondering about our tracking system. We recently learned from a colleague of yours, Bob Haskins, that Data Services, Inc. offers an excellent software package compatible with our computer setup.
Background information Request	We use a PXD 486 system with 8MB of memory. We are using a tracking system one of our employees formulated for our computer. We simply would like to expand our capabilities. Would you please send me information on your software package, with prices and installation guidelines?
	Thank you for taking the time to process this request. We look forward to hearing from you.
Close	Sincerely,
Sender's name, title address	Richard Bates, President Ship With Us 1245 Market Street Shawnee, Oklahoma 59869

FIGURE 2.3 A Block Format Letter

Now, evaluate the same letter (Figure 2.4) in *modified block* style:

Date	January 12, 1999
Inside Address Note the state is fully spelled	Ms. Martha Andrews, Manager Data Services, Inc. 2300 Broadway Ave. Louisville, Kentucky 40272
Greeting	Dear Ms. Andrews:
Purpose of the document	Recently our new company, Ship With Us, began providing freight, shipping, and hauling services to the residents of western Oklahoma. While we believe our services are as prompt and efficient as that of anyone in the field, the first six months of business has us wondering about our tracking system. We recently learned from a colleague of yours, Bob Haskins, that Data Services, Inc. offers an excellent software package compatible with our computer setup.
Background information	We use a PXD 486 system with 8 MB of memory. We are using a tracking system one of our employees formulated for our computer. We simply would like to expand our capabilities.
Request	Would you please send me information on your software package, with prices and installation guidelines?
Close	Thank you for taking the time to process this request. We look forward to hearing from you. Sincerely,
Sender's name, title address	Richard Bates, President Ship With Us 1245 Market Street Shawnee, Oklahoma 59869

FIGURE 2.4 Modified Block Format

All business letters, regardless of style, incorporate the following elements:

- *Date.* Always place the correct date on a letter. This provides a chronological record of events and allows you to verify both the date of sending and receiving important correspondence.
- *Inside Address.* The inside address reflects the name, title, and address of the person to whom you are sending the letter. Always try to ascertain the specific name and title of the recipient if at all possible. Verify the spelling of the recipient's name.
- *Greeting.* Use the greeting ("Dear") to address the recipient. Always use the title or the generic Mr. or Ms. of the recipient unless you know that person well and can comfortably communicate on a first name basis.
- *Attention Line.* If you are uncertain of the gender of the recipient, use a title (Dr., Sgt., etc). If you cannot determine the name or gender of the recipient (when, for example, you are writing to a number of companies for information), use an attention line in place of the greeting. An attention line (abbreviated ATTN:) allows you to address a specific individual through a job title as in this example, ATTN: Personnel Manager.
- *Letter Text.* The letter text is your opportunity to present your point, make your scenario, indicate good or bad news, or request action. Keep letter paragraphs relatively brief (three to six sentences), and remember to get to the point quickly.
- *Closing.* The standard closing for business letters is typically "Sincerely," though others are used. Avoid closings that are overly personal (for example, "Very truly yours," "Fondly," "Yours," etc.).
- *Signature Line.* The signature line comes four spaces below the closing regardless of letter style. Type your name, title, address, phone, fax, e-mail address, or any other pertinent information. Sign your name in the space between the closing and the signature line.

While the standard elements of business letters are the same regardless of letter style, do note these differences between *block* and *modified block* style.

- In block style, keep all lines flush left on the paper. In modified block style, indent the date, closing, and signature line to just past the center point of the page.

- In block style, double space between paragraphs and do not indent paragraphs. In modified block style, do not double space between paragraphs, but always indent paragraphs.

The choice of letter style is entirely up to you. Many prefer block style simply because all elements are kept flush left on the page.

TIP

Busy people who receive many letters in a day may not read each one carefully; therefore, make your point early in the letter and emphasize clarity, efficiency, and clear expressions.

Now that we've covered letter style and the standard elements in all business letters, consider, in addition, the types of letters typically written in business. You may write letters requesting action or services, you may write letters that reveal to the recipient either good or bad news, or you may write letters seeking information. Regardless of your purpose, most business letters fall into three distinct categories: action requests, good news, and bad news.

Action Requests

A large percentage of the correspondence that is sent back and forth between businesses constitutes action requests. Such letters might include requests for the following:

- information (for example, a letter seeking guidelines for a grant or proposal)
- specific action (for example, a letter inquiring about a job, requesting goods or services, or asking for help)
- consideration (for example, a letter proposing the feasibility of a plan, suggesting improved methodologies, or arguing for change).

All of these letter types seek something; the writer is actively pursuing either information or action. Thus, when you write an action request letter, remember to state your purpose clearly in the first paragraph, use a professional, courteous tone, and most importantly, keep the letter text clear, efficient, and to the point. Consider the action request letter in Figure 2.5 on page 34.

Date	February 24, 1995
Inside address	Dr. Marsha Bowen, Director of Graduate Studies North Hampton College Boston, Massachusetts 58960
Greeting	Dear Dr. Bowen:
Introduction of purpose Request for action	I recently learned of your new graduate program while reading the current issue of *The American Writer's Review.* I am particularly interested in the Master of Fine Arts program because I hope to pursue a career in writing. Would you please send me information on the program at the address listed below?
Specific information concerning the request	I am particularly interested in your program, which has a liberal residency requirement, a very important feature because I teach full time at Beechnut High School in Charleston, South Carolina. My schedule would permit me to attend classes only in the summertime, so your program requiring only one month of residency each year would be perfect for me.
	I am also interested in any information you could provide on financial aid, loans, scholarships, or summer assistantships. Thank you for your help. I look forward to hearing from you.
Closing	Sincerely, Thomas Marton 34938 Southville, #3 Charleston, South Carolina 38272

FIGURE 2.5 Sample Action Request Letter

Notice that the action request letter gets to the point quickly and addresses the recipient in a professional and courteous manner. Other types of workplace letters, however, require different considerations. Sometimes, you will write letters that contain good news for the recipient.

Good News Letters

Good news letters are items of correspondence that inform individuals of either a positive result in a previously difficult or questionable situation or of the successful conclusion of a project. These types of letters are somewhat different than the others we have discussed in that "good news" is typically revealed very early in the letter, before any explanation of *why* the positive outcome occurred. Consider the good news letter in Figure 2.6.

Date	July 18, 1996
Inside address	Mr. Charles Harrison 4876 Bluff Street Boulder, Colorado 29127
Greeting	Dear Mr. Harrison:
Notification of the good news	We here at ISIX, Inc. pride ourselves not only on the quality of our software products but also in the faith our repeat customers have in us. As a result, we are happy to refund, in full, the $453.95 that you spent on our DesignDraw program in March 1996. We are sorry that the program never worked properly for you and can only guess there was a flaw in manufacturing.
Background information	We have received no other claims for refund on this particular product, so we hope that yours was an isolated situation. In addition to the full refund, we are sending you a $50.00 gift certificate for use toward any other ISIX product. Our customers are important to us, so we will carefully examine the damaged product you have returned. We apologize for any inconvenience you have experienced as a result of this situation and hope you will consider purchasing a replacement DesignDraw program, one of our most popular products.
Closing	Sincerely,
Signature line	Rick Aston
Sender's address	Vice President, ISIX Inc. 4716 South Summit St. Oakland, CA 95736

FIGURE 2.6 Sample Good News Letter

Notice that in the good news letter, the writer communicates positive results immediately, saving the explanation for later in the letter. Unfortunately, though, you will sometimes write letters that communicate negative results. Bad news letters are composed a little differently.

Bad News Letters

In the workplace, you may occasionally be called upon to write correspondence that reveals a negative result or unfavorable findings. Bad news letters are rarely pleasant to write but are necessary. Unlike action request letters in which your point is expressed early and clearly or good news letters in which a favorable outcome is addressed immediately, bad news letters require you to use a somewhat more indirect method of revealing negative information. Rather than opening your letter on a negative point, you instead begin by explaining your rationale for the bad news before you reveal it. Consider the example in Figure 2.7.

Date

June 19, 1994

Inside address

Ms. Diane Russell
3923 Fourth Ave.
Detroit, Michigan 94746

Greeting

Dear Ms. Russell:

Purpose of the document

When we here at FORM-Tech decided to hire a new Sales Operation Manager, we were not prepared for the number of highly qualified individuals who would apply. After reviewing over sixty applications, we narrowed the field to five exceptional candidates. You were one of those candidates because of your fine credentials and excellent work record.

FIGURE 2.7 Sample Bad News Letter

Additional background

Because we are able to interview only two of the remaining five applicants, we had the difficult job of narrowing the field one more time. We carefully reviewed each candidate's education, work experience, and special skills. Each potentially brought varied and worthwhile qualities to our company. We had to find some way to narrow an impressive field.

Bad news

We decided, therefore, to select those applicants who had specific job-related experience in field management, because the new Sales Operation Manager at FORM-Tech will spend a good deal of time working with our representatives at satellite offices. As a result, we must regretfully inform you that we will not be able to offer you the position.

We commend you on an excellent resume and dossier and wish you continued success in the future.

Thank you very much for your interest in FORM-Tech.

Closing

Sincerely,

Signature line

Sender's address

Margaret Findlay
Vice President, FORM-Tech
8582 W. Front Street
Lansing, Michigan 94722

FIGURE 2.7 Sample Bad News Letter *continued*

Notice that in the bad news letter, the writer first makes a scenario for rejecting the applicant before revealing the "bad news." This method of making the scenario before coming to the point, eases the blow of negative information and allows you, as the writer, to maintain a professional and courteous demeanor.

SOME FINAL NOTES ON MEMOS AND LETTERS

Both forms of workplace correspondence, memos and letters, are vital communication tools in the workplace. Although increased use of e-mail and fax machines has changed the methods of workplace communication, memos and letters remain viable forms of correspondence because they allow you to form an important "paper trail" for your business and a more personal connection with the recipient. Remember a few final notes on memos and letters:

- Memos that you initial and letters that you sign confirm your complete understanding of and confidence in the contents. Never sign or initial a document that contains information you have not checked for accuracy. Signed documentation can be used in legal matters, so remember that what you sign, you endorse.
- Memos and letters constitute a "paper trail" in lengthy negotiations or complex scenarios. Send copies to relevant parties using a "cc:" at the bottom of correspondence, and keep a copy for your own files.
- *Draft* key memos and letters just as you would reports or other important workplace documentation. Revise and proofread correspondence before mailing, and importantly, do not mail your correspondence until you have carefully reviewed it, considered the consequences of it, and made pertinent changes.
- Guidelines for memo and letter writing are not restricted to just one type of correspondence. You may be called to write a good news or bad news memo or a letter-type report. The purpose in presenting each separately is to indicate that more often than not you will write the types of correspondence described in this chapter. Unusual situations, though, require critical thinking on your part. Carefully consider all the ramifications of your situation before you settle on either the type of correspondence you will write or the angle you will take.

Exercises for Classroom Discussion

Directions: The following mini scenarios are a means for you to practice what you know about memos and letters. In each situation, you will receive enough information to help you decide (a) whether to write a memo or letter and (b) what type of memo or letter to write. Write correspondence in response to the following five scenarios. At the conclusion of each, justify your choice of correspondence and correspondence type. (Alternate—as a class, discuss what type of correspondence you would write for each scenario, and justify your choices.)

1. Bob Becker is angry. As Personnel Manager, he has tried repeatedly to get the employees at Koller Construction Company to refrain from using the photocopy machine for personal copies, yet the copy budget, which he is in charge of monitoring, is consistently over budget by several hundred dollars. As the Assistant Personnel Manager, you have been given the task of communicating Bob Becker's displeasure to your colleagues. In addition, Becker has asked you to come up with some kind of plan that will slow the number of personal copies made on the photocopy machine. Thus, you must come up with a brief, feasible plan for curbing personal copy-making and communicate it to your colleagues.

2. You are the Parts and Services Manager at Fledbig Ford, a dealership in Pontiac, Michigan. Felicia Wills, who bought her car at your dealership, returned the car after three months complaining of tiny paint bubbles on the hood of her red Ford Mustang. You and an assistant carefully studied the paint bubbles to determine whether or not they were flaws in the manufacturer's paint job or rock chips. After careful consideration, you decide that the manufacturer is responsible for the tiny air bubbles on Ms. Wills' hood. You must write to let her know that you will repaint the hood at no charge.

3. You own a home in a quiet, small community of roughly fifty year-round residents in rural Wisconsin. Because the community is situated on a large stretch of "sandy loam" type soil, a race track developer, Jack Grange, has made an offer on a ten-acre parcel about two miles from your home because he believes the land to be perfect for a stock car race track. You and your neighbors are horrified at the possible intrusion on your quiet community. First, the neighborhood is accessible by just three poorly maintained gravel roads; the increased traffic would certainly deteriorate the roads. Second, most of the home owners are retirees and have no interest in stock car racing. Third, home values, you believe, could plummet should the race track be established. You decide, as a result, to write to your state senator, Barton Styles, to try to stop this purchase.

4. You are a newlywed planning to take your spouse on a late honeymoon to Honolulu, Hawaii. You need to know about the city, hotel availability, travel, sight seeing, and, perhaps most importantly, general accessibility for the handicapped (your spouse is wheelchair bound).

5. You are the Assistant General Manager at a small manufacturing company that makes lamp shades. For years, the leftover heavy cardboard your employees use to make some lampshade frames has been burned in the company incinerator. Because the company is small, the amount of burning has not been significant, but recently city officials wrote to you indicating that tough new pollution laws prevent even minimal burning at a company the size of yours. As a result, you were given the task of implementing a new recycling program. Instead of dumping leftover cardboard into large bins for burning, the company must begin separating types of cardboard (according to weight) and other paper products (correspondence, paper toweling, etc.) into designated bins. You must not only implement the new plan, but you must also inform your workers of the changes.

SCENARIOS FOR MEMORANDA AND LETTERS

Accounting for New Computers

CONSIDERATIONS

In this scenario, you are interested in obtaining new computer equipment for your company. You experience, however, some drawbacks in this project. While reading through this situation, keep in mind the following questions.

- How well do your co-workers understand the importance of new computers?
- What are the political and social dynamics between you and Mary Fenton?
- What outcomes are more important to you—the quality of the computers or the professional relationships involved?

You have recently moved to Peshtigo, Wisconsin from Chicago, Illinois. Peshtigo is a small northern town with a quiet, cozy downtown business district, several rivers and lakes, and a great school system for your children. The air seems cleaner in this town, the local fanaticism with the Green Bay football team is contagious, and most of all, working for a small accounting firm has been a refreshing change of pace. The stresses of your former life in Chicago seem far away, although you miss some of the energy of the big city.

One aspect of the new job, however, has been troublesome. You were accustomed to working on better quality computers than your firm, Anderson Accounting, has in its office. Their machines are so antiquated that you've resorted to bringing your personal laptop to work—after all, the more clients you can serve, the more money you make.

Today, while looking up an obscure tax code in the firm's resource room, the president of the company, Arlice Anderson, walks in and smiles at you as you look up from a thick book.

"I've been looking for you," he says.

"Here I am," you answer. You like "Anderson" as everyone calls him. He recruited you after watching your presentation at a tax accountant's conference in Chicago.

"I've noticed that nice computer on your desk," he says, taking a chair across the table from you. "Yours from home, I suppose?"

"Yes," you reply. "Is that a problem? I don't save any client information on it. In fact, I move everything to a floppy disk. It's easier to use the newer programs."

"Oh, I wasn't worried about confidentiality," Anderson says. "I just wondered if you knew enough about computers to find some good ones for the whole office. It's time we bought newer equipment, and we had a good year last year, so there's money—about $30,000 in the equipment upgrade account. We've been putting this off for a while."

"I'd be happy to do the research," you say. "What kind of computers were you thinking about buying?"

Anderson laughs heartily. "I know practically nothing about them. In fact, we'll probably have to bring someone in to teach all of us. But I do want those little ones that you can carry around, like what you have."

"They're called laptops. You know, one of my new clients is a company that teaches people how to use computers, and they sell them, too. Maybe we can work out a deal with them."

"Paul Harris at Computing Essentials, right?" Anderson says. "We could trade for services."

Back in your office, you start making a list of the firm's computer needs.

Secretarial—two computers to do word processing and scheduling.

Accountants—three fast computers to handle tax returns and bookkeeping. Internet connection? Need CD-ROMs for tax-code reference disks.

Management—one fast computer with Internet access and CD-ROMs for reference material disks. Anderson wants a laptop for himself.

After calling Computing Essentials, the company you mentioned to Anderson, you are told the president, Paul Harris, a young man in his 20s with enormous energy and a great personality, was out of the office. You decide to fax your list to the secretary, so Paul could come up with some ballpark figures when he calls you back.

To your surprise, one of the secretaries brings a fax from Computing Essentials to you only two hours later. It reads:

Sorry for the briefness of this fax, but I will be out of the office this afternoon for a service call and wanted to give you something to think about today. Here's what I have for you:

2 Ellitech MaxPro 2000 Desktop Computers @ $2,999 each = $5,993

400-MHz processors, 64 MB RAM, 4-gig EDO hard drive, 17-inch monitor, 24X CD-ROM drive, 3.5-inch floppy drive, fax/modem, keyboard, speakers, mouse. Comes with our Level One software package installed.

3 MobileMax 28000 Laptop Computers @ $3,899 each = $11,697

400-MHz processor, 64 MB RAM, 3.2-gig EDO hard drive, active matrix display, 6X CD-ROM drive (swap with floppy), 3.5-inch floppy drive (swap with CD-ROM), PCMCIA 36k modem, touchpad mouse, power cord. Car adapter and extra battery available for $600 each in addition to the price listed. Comes with our Level One software package installed.

1 Ellitech Performance 100 Laptop with Docking Station @ $4,255 each = $4,255

400-MHz processor, 96 MB RAM, 4-gig hard drive, active matrix display, 24X CD-ROM drive, external 3.5-inch floppy, trackball mouse, power cord, carrying case. Docking station has a 2-gig hard drive, internal 3.5-inch floppy drive, and internal 56k fax/modem. It also comes with an Ellitech inkjet printer (very small and portable), which prints 360 dpi black-and-white at about 3 pages per minute. Comes with our Level One software package installed.

Total estimate: $21,945

NOTE: Level One software package includes the latest version of the Windows operating system and a budget of $300 per computer to choose which software titles you would like to purchase from our list.

All equipment comes with a three-year manufacturer's warranty against defective parts. You may also enter into a service contract with us for $300 per year to take care of repairs and upgrades for up to five years. This contract covers labor only, not parts (most parts covered by the manufacturer's warranty for the first three years).

continued

I'm guessing that you may need some training, and I'm willing to do group training for your company at $50/hour, which we can trade for the accounting services you're providing us.

I'll call you first thing in the morning to discuss this quote.

Thanks for thinking of us,

Paul Harris

Rubbing your forehead, you pause to think about what most of this jargon means. The processors make the computer run faster. Your laptop is a 233 MHz, so a 400 or higher must be better. You also remember needing more RAM so the computer could handle the large amount of number crunching required in accounting programs—64 seems like a decent amount. As for hard drive space, that's easy—the bigger the number, the better. You're not sure about modems, though, and what is a PCMCIA? Researching these things so you can explain them to Bob as part of your equipment proposal information becomes your top priority.

Another accountant, Mary Fenton, pokes her head in your office door. "Heard you were looking for new computers for everyone," she said.

"The rumor is true," you answer. You're not sure about Mary just yet—she's been a little hard to read from day one. "I think the accountants and Anderson will get laptops, and the secretaries will have desktops. I just got a fax from a company, and we're looking at a total of around twenty thousand dollars right now."

"That seems high," Mary says, frowning.

"Not really. Laptops are more expensive, but that's what Anderson wanted. I'm assuming he thinks it will work better for us when we're visiting clients."

Mary looks over your shoulder at the quote and says, "Wait a minute. I think I saw something better in a catalog. Let me get it."

She returns in less than a minute with a computer catalog and triumphantly plops it on your desk. "See," she says, pointing to one item. "These are only twenty-two hundred dollars."

You examine the specifications printed next to the laptop computer, a "MaxTechnia Pro."

Muttering to yourself more than to Mary, you read them off, "Two hundred megahertz processor, eight megabytes RAM, one gigabyte hard drive, dual scan display, floppy drive, trackball pointing device, battery, power cord, and carrying case. . . . It doesn't mention a warranty or CD-ROM anywhere, or software."

"We can buy those CD-ROMs that hook up to the computer and share it. What do you think?" Mary says.

"I don't know," you answer. "These don't have the same components as the ones from Computing Essentials. I'll have to ask Paul about it."

"He'll just tell you his are a better deal," Mary said.

"Of course," you answer, "but I'm sure he'll tell me what the difference is. He's an honest guy."

After Mary leaves your office, you decide to call Computing Essentials, hoping that someone can answer your questions. Paul's assistant, Kristin, answers the phone and assures you she can help.

"Well, the biggest difference is the display," she says, after you read off the features of the MaxTechnia Pro. "Active matrix displays have a much clearer screen, and they look more like a regular monitor does. Dual scan displays are harder to see from different angles, which means you have to stay very still or keep adjusting the screen position. I've found that they cause more eye strain, because many are dimmer than active matrix and not as clear."

"What about the difference in processor speed?" you ask.

"Well, manufacturers are not making two hundred megahertz processors any more because they don't work as well with the newer, larger software programs. I'd trust their instincts and go with a faster processor. Oh, and I'm really surprised at how little RAM they're giving you. Laptops need more RAM than desktops, and you need at least 24MB to run the newer software programs."

"Does the smaller hard drive space mean anything?" you want to know.

"The laptops for the accountants won't have docking stations, right?" she asks.

"Right."

"Well, then you'll need more storage space. You can take up one gig just with programs since they're getting larger and larger. I think with one gig you'll be saving to floppy almost all the time, and that can be a hassle. Floppies can be pretty unstable too—they can lose data under certain circumstances."

"Such as?" you ask, thinking about all the data you have on your floppies.

"Oh, exposure to magnetism from radios, manufacturer's defects, getting scratched or bent. Things like that. We've had people claim that bringing them in from the cold and putting them in a warm computer can corrupt the data as well."

"Great," you say. "I believe it gets pretty cold here?"

"Oh yeah," Kristin says, laughing.

"Thanks for your help," you say. After you hang up the phone, its intercom buzzes. It's Anderson.

"How are you coming on the computer quote?" he asks. "Mary was just in here and mentioned you were looking at some pretty expensive machines."

You find yourself grinding your teeth, and realize you haven't done that since leaving Chicago. "Fine," you answer. "In fact, I'll write up some of the information for you so you can see the direction I think we should go. I'll have it on your desk before five."

"Great," Anderson says. "Did Harris want to trade for training us?"

"He sure did."

"Good. I look forward to seeing what you've come up with." You hear the intercom click as he disconnects and turn to your laptop to start writing. You wonder what's going on with Mary, and whether or not this penny-pinching is something the company encourages. After all, you haven't been here that long, and you don't know how much influence Mary has with Anderson. You think you've done well because almost $9,000 will be left in the equipment upgrade budget, but you haven't even started investigating printers. You may or may not have to replace the ones already in use. You sigh and begin typing, glancing at the clock to see if you have enough time to think this through well before five. You still have to log on to the Internet and find some good, simple explanations for some of the components you're unsure of. You don't want anyone to have big questions or be uniformed; after all, whatever computers Anderson buys, you'll probably have to live with for many years to come. You hope your write-up of the computers from Harris' company shows Bob that Mary's opinion is off-base.

Suggested Next Steps:

- Make sure *you* understand all the computing terminology. Consider Anderson's level of understanding and any possible questions he may ask you. The Internet may be a good source of basic information on computer components and their functions.
- Decide how you want to write up this information to Anderson, keeping in mind that he may or may not be the only person to read it. How much detail and explanation do you want to include?

- Write to Anderson, detailing the results of your research into new computer purchases, and include any other information you feel is relevant.

The Archaic Report Form

CONSIDERATIONS

In the following situation, you must revise an ineffective document; however, doing so might offend your boss. While reading the scenario, keep in mind the following:

- If you were investigating a burglary, what pieces of information would you need from the crime scene, witnesses, and so on?
- In this situation, what importance is placed on paperwork by the police officers, the sergeant, and perhaps other readers such as lawyers or insurance agents?
- What are the political (career-related) issues in this situation?

It's Monday morning. You've just arrived at the Sheriff's Department in a small, rural Midwest town where you are a new patrol officer. You've been on the job for three months, and while you like the freedom of the work, driving around the country roads to which you have been assigned, you've noticed a few things about the work environment that could be streamlined and improved. You haven't really made many friends at the department yet, though you feel comfortable talking to a few of the other officers.

"Hey, Bob," you say as a fellow officer walks by on the way to his desk. "That was some crime scene you were on yesterday out on Brooksville Road. A bunch of crazies burglarized and vandalized the Smith's camp, eh?"

"Yeah," says Bob. "Luckily, no one was at the place. They could have been hurt."

Bob seems distracted to you, so you ask him if anything is bothering him.

"Oh, nothing," Bob answers. "I'm just pretty fed up with the forms I have to fill out. I dread this report. There's gotta be a better way."

You can understand how Bob feels. Last month, when you were very new to the job, you faced your first burglary report. A couple of drifters decided to break into a farmhouse owned by an elderly couple. The suspects didn't steal much and didn't get far before a state police patrol picked them up driving the old man's truck. The report that followed, however, was time-consuming. You had been warned about all the paperwork associated with police work, but the burglary form was practically archaic and required information that seemed irrelevant or ignored information that seemed to you to be important. You didn't say much, though, because others told you the patrol sergeant didn't like change and certainly didn't like problems in his relatively small operation.

"You think maybe Sgt. Richmond might consider a rewrite of the burglary form if I offer to do it?" you ask Bob. "I mean, I'm not saying I can do that much better or anything, but it took me a couple of hours just to figure out *how* to fill it out when I had an incident last month."

"Tell me about it," said Bob. "This is probably the hundredth time I've filled out one form or another, and I still hate it. Sure, go ahead and deal with Richmond if you feel like making waves. I heard that Chief Langley is kind of down on Richmond right now because his paperwork for a manslaughter case didn't look too sharp."

"I'm no writer," you respond, "but I can make some suggestions, I think."

"Look, pal, don't take this the wrong way, but Richmond really dislikes new guys with new ideas," Bob gently warns him. "He's been sergeant here as far back as anyone can remember. You better have some good ideas in writing if you want to get to him."

You really feel torn in two directions here. You do have some good ideas about improving the report form, but you don't know how to explain those ideas to Sgt. Richmond so that he'll understand that you just want to make life easier for everyone. You go to the desk secretary and ask for a copy of the burglary report form, which looks like this:

Burglary Report	
Location of Occurrence	Victim Name(s)
Point Where Entrance Was Made	Implements Used to Make Entry
Location of Occupants During Occurrence	Date/Time of Occurrence
	Date/Time Reported

continued

How Many Suspects?	Sex and Descent
What Was Taken?	Trademarks of Suspects
Type of Premises Entered	Suspect Vehicle (if vehicle used)
Units Responded, Investigative Div. Notified	Victim Occupation & Descent
Witness Information	Victim Address
Victim Phone	
Suspect List and Information	Crime Reconstruction

You decide to take the old burglary report home and write up some suggestions for improving it. *Then what?* you think. *How do I explain the changes to Richmond, explain why the station should revise the form, and explain that I'm just trying to improve productivity on the job? How do I keep the sarge from getting angry at me?* You decide that a face-to-face meeting might be a little confrontational, so you'll just put your suggestions in the sergeant's mailbox. Maybe it will give him time to think about your proposal before you talk.

SUGGESTED NEXT STEPS:

- Study the old burglary report and think about the kinds of information you would, logically, need to investigate and perhaps catch the burglar. Remember that the victim is often too shaken to offer detailed information on his or her own, so try to include items in the form that will draw as much relevant information as possible out of the victim or any witnesses.

- Decide which revisions would be the most important. Then decide which revisions could wait but are still important changes that you believe should be made. Lastly, you may have some ideas for revision that are not crucial but would be helpful to officers working with the form.

- Taking into consideration the information you have about the sergeant's attitude, write to him about your suggestions for revising the burglary report form.

The Air Bag Problem

CONSIDERATIONS

In this scenario, you tackle the not-so-pleasant task of sending negative news to your superiors. As you read the information, keep in mind the following:

- What problems can you identify concerning technology and corporate issues in this situation?
- Examine the technical data. How can you best represent this information clearly?
- What are the ethical implications involved in the way you send this negative news?

You walk into the crash test laboratories of Rhinex Testing, Inc., a company that has employed you as an automotive testing engineer for five years. Most of your time is spent evaluating the safety features of contemporary automobiles, including built-in child restraint seats, adjustable safety belts, improved auto safety glass, and reinforced bumpers, for one of the country's top three automotive manufacturers. You like your job and enjoy crash-test trials, and with two children of your own, you relish the opportunity to play a role in safer driving.

After pouring what will be the first of several cups of coffee, you notice your colleague, Marta Richmond, frowning at the microwave oven.

"What's wrong, Marta?" you ask. "Having trouble nuking that hot chocolate?"

"No," she replies smiling. "I'm not frowning at the microwave oven. I'm frowning at the results John gave me from yesterday's testing on the new air bag systems. Have you seen the preliminary report?"

"I really don't know much about John's work on that, but I did some air bag testing myself about a year ago. What's the problem, anyway? Air bags are mandatory equipment now, and there have been modifications to reduce the inflation speeds."

"Yeah, I know, but the crash testing John did on infant and kid dummies was not good," Marta responds. "Three out of the five tests John did yesterday on the infant and small kid dummies pointed to their death from suffocation or broken necks in low-speed crashes. The air bags still inflate so fast that the dummies were wrenched backwards with incredible force."

"You know that when I did air bag tests, we only crash-tested adult dummies," you reply.

"Well, most people keep their kids in the backseat, but a recent case of a child death in Denver caused the manufacturer to contact us for testing. There's been some newspaper reports that three child deaths have been attributed to the newer air bags. We've tested at car speeds ranging from five to thirty-five miles per hour with infant and kid dummies from two to eight years of age. The six-to-eight-year-old dummies were injured regardless of impact speed. The infant and younger kid dummies, though, were killed nearly every time, again regardless of speed."

"Wow," you respond to the numbers. "The home company needs to know about this, and they aren't going to like it."

"Tell me about it," Marta quickly adds. "John and I have to report our findings. Fortunately these new, slower-inflating air bags haven't been installed in any cars available to the public. Still, the news isn't good. The big guys were counting on these new bags as a public relations feature of the new line. It wouldn't be so bad, but we're both relatively new here, and we don't begin to know how to report such bad news. This could mean a delay in shipping until the problem is solved."

You consider the possibilities. Marta is right that Marshal Motors, Rhinex's parent company, won't like the findings she and John put together. It's also true that John and Marta are new and could use a little help dealing with the "big guns" who won't be happy.

"I'll tell you what, Marta. I've had to deal with bad news before. Remember when the side-mounted gas tanks caused so many problems? I wrote that letter and a follow-up report. I'll help you and John write this letter. I think I have enough information to draft a letter that the two of you can then review and revise. What do you think?"

"Oh, God, thanks," Marta exclaims. "We were so worried about that write-up. What can we do to help? What else do you need to know to write the letter?"

You stop for a moment and think. "Well, I need to know the models you tested. Also, is the inflation speed the same on all models?"

"Yep," Marta quickly responds. She pulls out a sheet of paper, obviously her summary notes from the tests. They read:

Slow-Inflating Air Bag Test Results
—infant tested in both front and rear-facing car seats
—2–5 yrs tested in front-facing car seat

Excursion XRZ four-wheel-drive sport utility

Speed	Dummy Type	Injury Rating
20 mph	infant	fatal
	2–5 yrs	critical

Illusion LF four-door luxury model

Speed	Dummy Type	Injury Rating
20 mph	infant	critical
	2–5 yrs	critical

Roundabout two-door, compact style

Speed	Dummy Type	Injury Rating
20 mph	infant	fatal
	2–5 yrs	fatal

Nadir 100 mid-size, two-door family model

Speed	Dummy Type	Injury Rating
20 mph	infant	fatal
	2–5 yrs	critical

Marta sees that you're done reading and says, "The vehicle didn't seem to matter as much as the impact speed and the age of the child. Like I said, all the infant and small kid dummies died or were injured in crashes of even twenty miles per hour. The bags are so large, they virtually suffocate the child."

"Does this mean the bags are worthless?" you question.

"Well, not necessarily," Marta quickly answers. "We can recommend that parents put their children in the backseat. Problem solved. Not all parents will do that, I realize. Also, the bags appear to be fine for adults."

"Yeah, but the bags sound like a problem for kids," you add. "This letter is going to take a little diplomacy."

You take away the information you have and draft a letter to the home office, Marshall Motors, Inc. You'll probably write it to the Senior Vice President for Manufacturing, Stephen Youngs, and copy it to the Safety Manager, Roy Nishi, a good friend of yours. Marshall recently settled a lawsuit with a driver over faulty rear hatch latches. Youngs will not be happy about another manufacturing or engineering glitch. You definitely want to communicate the bad news, but you also want to try and find something positive in the situation. This one won't be easy.

SUGGESTED NEXT STEPS:

- Study the information Marta provided you, and think about what this information means in terms of how it affects the way you will communicate the results to management.
- You may want to do some outside research, such as finding newspaper or magazine articles about child/infant injuries and deaths attributed to air bags.
- Write to the appropriate people, as mentioned in the case, communicating the test results.

Sunset Inn Accessibility

CONSIDERATIONS

In this scenario, you are asked to incur extra expenses, at a time when you are already losing business while you renovate your facility. When reading the next situation, keep the following in mind:

- What do you hope to gain by renovating your hotel?
- What are the political factors involved with handicapped accessibility?
- What other people or organizations have a stake in the success of your business? How are these entities affected by your request for additional funds?

You are the owner and manager of Sunset Inn, a small inn on the shores of Lake Peaceful in northern Montana. Most of the inn's business occurs from late June to early September, during the region's nicest weather.

The Sunset Inn offers quiet, smoke-free rooms, all with views of the lake, and is very close to a large state park. Most guests stay for a day or two, tour the park, walk the nearby nature trails, and eat in local restaurants. Most of the guests are outdoors-type people. The town is small and offers a few restaurants, two gift shops, and a very small historical museum. The town's main attraction is the state park, several waterfalls, and the many miles of white sand beach.

The inn has 30 rooms with balconies off each room on the second floor. Sunset Inn also houses its own laundry facilities, where the four housekeepers wash the bedding and towels daily. Your other staff includes two clerks, a night auditor, and an assistant manager.

The inn does a brisk business, and you have just received a $200,000 loan from the Montcalm Bank and Trust to update the rooms with new trim, wallpaper, paint, furniture, and bathroom fixtures. The contractors will begin work at the end of the season.

You open today's mail, surprised to see a letter from the Northern Montana Association for the Disabled.

The letter reads:

Northern Montana Association for the Disabled
1010 South First Street
Big Valley, Montana 59822

The Sunset Inn
1010 Lake Drive
Big Valley, Montana 59822
June 1, 2001

Dear Manager/Owner:

It has come to our attention that your inn does not contain a room designed for guests requiring wheelchairs. The technology and fixtures to accommodate *all* guests is affordable and readily available. We ask that you consider renovating one of your first-floor rooms with a special shower and lower counters, beds, shelves, dresser, night stands, and door locks.

While you are not required, under the Americans with Disabilities Act, to provide such accommodations (because your building is an older structure), our organization will list your inn in our yearly recreation guide for disabled people if you do. We hope you believe, as we do, that everyone should be able to travel and find comfortable lodging in every town in the our region. We are counting on you to provide such comforts to the disabled, should you make the suggested changes.

Sincerely,

J. Swanson

Jessica Swanson, Chair
Northern Montana Association for the Disabled

She has a point, you think after reading the letter. *We should have appropriate lodging for guests with disabilities. I should find out how much this would cost and write to the bank.* Your loan officer is Carol Horn, a friendly but scrupulous woman who wants you to keep to a strict budget. However, you might borrow some more money from the bank if you provide a detailed breakdown of what it would cost to make a room handicapped-accessible. You start jotting down notes on the items you will need—a lower dresser and bathroom sink/counter unit, bed, and door locks, and that special shower. You guess that you might be able to find some of this information through the Internet or by calling a local contractor. Excited about the project, you think about how to approach Carol with your request.

Suggested Next Steps:

- Determine the cost of the handicapped-accessible items mentioned in the case. You might find this information on the Internet, by contacting a local organization or government agency that assists the disabled, or from your instructor.
- Consider the objections your loan officer might have to increasing your loan, and how you might address these objections.
- Write to your bank with the intention of receiving more loan money for this project.

Soil Incineration: The Groundwork

Considerations

This is the first in a series of scenarios involving a proposed contaminated soil incinerator. You will need the information from this situation for the soil incineration scenarios in subsequent chapters. While reading "The Groundwork," keep these questions in mind:

- Why is your company proposing a soil incinerator? Of what benefit is the project?
- What are the potential drawbacks to the incinerator's location? What information are you given about the incinerator site?
- Who provides you with information? What stake or standpoint do these people seem to have concerning the project?

It's your first day working for Donnelly Engineering Company to prepare a proposal for the state's Department of Natural Resources. Donnelly engineers brought you in as a consultant because of the sensitive nature of this particular project.

Jack MacGregor is your prime source of information and guidance as you familiarize yourself with the project—a proposal to build a soil incineration plant. Jack is the head of Donnelly's engineering team and has scheduled a meeting for the two of you today.

You walk into the Donnelly meeting room, a rather small place with floor-to-ceiling windows along one wall. A triangular, glossy black-topped table surrounded by seven gray, cushioned swivel-chairs dominate the room. Jack is already seated near the apex of the triangle and motions you to sit next to him.

"Good morning," you say.

"Morning," he answers, hardly looking at you. He seems to have no papers with him, just a legal pad with some notes written on it. "I have a few things to relay to you before you start researching. You must be very careful as you gather information."

"Understood," you reply, pulling a legal pad out of your briefcase and preparing to take notes.

"First, we're proposing the soil incineration plant because of the great amount of contaminated soil stockpiled in our state and surrounding states. Environmental laws are making it harder to clean up after accidents or leaks from underground storage tanks. Most of the soil has gas or oil in it, but there's no place to put it or clean it up relatively inexpensively."

You nod, though you had already heard of this problem.

"So," Jack continues, "we approached a land owner outside of Cedars, Jeff Haywood, who is willing to let us lease some of his land. It's remote but has a good road and is within driving distance to several cities with rail service for transporting the soil."

"Sounds perfect," you say.

"We thought so. But there are five homes located within a ten-mile radius of the site. The homeowners caught wind of the project and don't like the idea."

"Why not?" you ask.

"Well, they're worried about the contaminated soil getting into the groundwater. We tried to explain that the contaminated soil is stored on an asphalt lot and is always covered. They also think the treated soil will have too many chemicals remaining in it. And there are some who don't want an industrial complex in that area. It's all farmland and swamp. Anyway, you need to be aware of all this so that when you start gathering information, you're prepared."

"Thank you," you say. "Perhaps people will relax when we receive the government okay for the plant. What kind of information does the Department of Natural Resources require to approve the project?"

"The basic run-down—a description of the process, a detailed description of the plant and its location, and plans for operation including our testing to comply with environmental regulations."

"I'll start with the plant site," you say. "I know people at the Soil Conservation Office so we can get that information quickly."

"Fine," Jack says, but he seems a little uncomfortable. You have no more questions for him, so you leave the meeting.

From past experience with this type of project, you know you must find out how many families live near the proposed incinerator, what use they are making of their land, the area's watershed system, and the type of soil designation. You write to Barbara Martin, head of the local Soil Conservation Office, for that information.

Three days after writing to Ms. Martin, you receive a letter in the mail.

We received your request for information and provide the following information:

Three other families live within ten miles of the site; all have dairy farms located off Tamarack Lane.

The Shewanee River, a seasonal home to geese and other protected bird species, runs within five miles of the property and is fed by a tributary that runs through the property.

According to our records, the land on which the incinerator is to be located has a soil designation of "sandy loam" and is rated as "extreme" for sanitary landfill use. Additionally, the land around the tributary, two acres from the proposed site, is classified as "swamp" and rated "extremely severe" for absorption.

I formally request that you call me at your earliest convenience to discuss this matter.

Sincerely,

Barbara Martin

You are concerned about the apparently poor conditions for the proposed site, and call Barbara immediately.

"I'm glad you got back to me right away," Barbara says. "I understand that Donnelly just brought you in on the project, and you don't know what's been going on."

"Going on?" you ask. "I don't understand."

"This incinerator is a hot topic in Cedars," Barbara continues. "Residents formed a citizen's group called POL for Protect Our Land. They think that contaminants will leach into the soil and eventually wash into the Shewanee. I thought you probably didn't know about this because I told Jack MacGregor over there that they couldn't pick a more absorbent soil area to put the incinerator. I also told him about POL."

"Well, thank you for the information," you say, not wanting to sound suspicious of your own company but also wanting to ask Jack why he didn't let you in on this during your meeting. "Of course, Donnelly is concerned about the environmental impact of the plant. We need the official report for our proposal."

"You're welcome," Barbara says. "Personally, I don't think it's a good place for an incinerator. You know, the soil coming out of those things still isn't very clean."

"Thank you for your concern," you say and end the conversation. After hanging up the phone, you sit and think for a while. Why didn't Jack tell you about POL and the soil designations? You had planned to write him and provide a summary of the Soil Conservation information so that he would see that your progress on the project is on track. Now it seems he already knows this information, but you decide to give him the benefit of the doubt. Maybe he didn't understand the impact of what Barbara told him.

SUGGESTED NEXT STEPS:

- Decide what pieces of information you should provide Jack. Defend your decisions to include or omit information.
- Write up a summary of the Soil Conservation information. In your Solution Defense, explain your strategy for organizing this information.
- Consider the implications of your summary on the future of the project. Discuss these considerations in your Solution Defense.

CHAPTER 3

Ethical Considerations and the Role of the Technical Communicator

Imagine the following scenario. You are a technical writer at a new computer company that specializes in creating web pages for clients in manufacturing, especially those that manufacture industrial implements. You and your best friend from college were amazingly fortunate to land jobs at the same company, doing what both of you love most to do—create and design web pages. Your boss is happy with your work, but he's happier with your partner's work because it is highly creative, pleasing to the client, and produced in nearly half the time you produce your web pages. You know why the work is completed so quickly.

Your best friend, your roommate through four years of college, is copying or completely lifting entire web page designs from other, smaller competitors. When you caution him that someday someone will notice the remarkable similarities among web pages, he chuckles, calls you paranoid, and reminds you that the Internet is a very, very big place with very little policing. He's not worried. Are you?

What would you do in a workplace situation that requires you to weigh, on one hand, your commitment to a friend, and on the other hand, the ethical "correctness" of that friend's activities? Some might be tempted to say, without considering the ramifications, that a friend's a friend; you never betray a friend. Others might say, though, that you'd be doing your friend a favor by leaking the information to your boss; plagiarism of any kind is serious. Regardless of your response to this scenario, recognize that whatever you decide is "right" in this circumstance you must face the ethical consequences of your decision.

ETHICAL DECISIONS ARE PART OF EVERY WRITING SITUATION

Remember, that much of what you do as both a technical communicator and a student using this text is to persuade. In fact, virtually every scenario is really an ethical situation because of the human factors involved. Ethics must be a part of every writing situation if we acknowledge that our job is often to convince others either to support our conclusions or to accept our judgment as correct. If you wish a project to be funded and you write a report to your boss, you are attempting to persuade. When you persuade, you must consider all the variables underlying that action. Even if you are only reporting information, you have an obligation to consider all the factors both implicit and explicit in your report. Considering both the ramifications and the results of a decision must be a part of the development of any ethical writer.

This chapter, therefore, comes early in the text because ethical considerations are a fundamental part of what you do as a technical communicator. Most dictionaries define ethics as *a belief in moral duty; moral principles, quality, or practice.* How, though, are issues of morality or duty relevant to your role as technical writer in the 1990s? This chapter should familiarize you with some trends in the area of ethics, and prepare you to address the difficult ethical situations that arise in the scenarios of this chapter.

WHAT ARE YOUR ETHICAL RESPONSIBILITIES?

In his article, "A Basic Unit on Ethics for Technical Communicators," Mike Markel notes that modern technical communicators

> . . . will have to make decisions about how to treat clients, customers, and other organizations. They will have to make decisions about how their organizations deal with government regulatory bodies. And they will have to make decisions about how the actions of their organizations affect the environment.*

Does this mean that the entire weight of decisions made by your company rests squarely on your shoulders when you write documents reflecting your company's practices? Obviously the entire weight doesn't rest on your shoulders alone, but some of it certainly does. Making decisions, therefore, based on solid reasoning and ethical principles should be an important component of your work. Unfortunately, making good decisions isn't always easy.

*Markel, M. 1994. A Basic Unit on Ethics for Technical Communication. In *Humanistic Aspects of Technical Communication*. ed. P. Dombrowski, Amityville, New York: Baywood Publishing Company. p. 199–221.

You cannot expect, for example, that everyone with whom you work feels the same way you do about anything. What you might regard as unethical, for example, the decision to deny an elderly woman medical benefits because her median income is roughly $200 over your cutoff number, might simply be regarded as a policy decision by someone else. In fact, are policy decisions of this sort inherently wrong? Is someone who makes a decision based strictly on policy avoiding the ethical consequences of decisions? To make the best ethical decisions possible and to respond as logically and reasonably as possible to the ethical scenarios that follow, you need a few pointers on how to sort out the problems inherent in ethical dilemmas. While there is rarely a right or wrong answer to ethical situations, there is usually a "right" answer for you that you will support in your Solution Defense. Formulating an answer means paying close attention to the site of the ethical problem, the implicit and explicit aspects of ethical scenarios, the ramifications of decisions, and the foundational questions that you must ask after you have come to an ethical conclusion.

LOCATING THE SITE OF THE ETHICAL PROBLEM

You will begin the process of responding to this chapter's scenarios by "locating the site" of the ethical problem. Locating the site means determining the specific aspect of any given scenario that forms the fundamental conflict in an otherwise routine situation. All technical writing situations contain some element of conflict. Some can be very difficult to resolve because of the *degree* of conflict. Your first step as a technical communicator is to sort out the players, the situation, and the conflict. Once you've done this, you can begin to formulate a response.

The Players

Perhaps the most important step in sorting out a solution to any ethical situation is first evaluating who is involved in the process. Isolating the players and their roles will help you determine who occupies a role of responsibility, who is subordinate, who might be affected by the decision you make, and who or what caused the conflict in the first place. Always begin the process of solving an ethical scenario by listing the people involved. List also the following information, if it is provided:

1. *Job Title.* What is the position of people in the company hierarchy? Who is the boss? How do their positions relate to yours (equal status, higher or lower on the corporate ladder)?
2. *Relationship to other people.* Who is employed by whom? Which people have work relationships that are competitive or downright ad-

versarial? Which people seem to work as a team or try to help each other as much as possible?

3. *Any discernable personality traits.* Who, in your opinion, has a tendency to be argumentative? Overly passive? Disinterested in the job? Always trying to please the boss at any cost? You may develop a perception of personality traits by what other people tell you. For example, one co-worker might comment, "Mary drinks a pot of coffee a day when she's stressed." Later, you observe Mary returning often to the company coffee pot. Can you theorize from your observations that Mary is currently experiencing stress?
4. *Habits.* Who has demonstrated trustworthiness? The propensity to divulge confidential information? Who regularly comes in late to work? Who stays after hours to finish projects? Who always makes excuses when things don't go right, and who are quick to blame themselves for errors that might not be entirely theirs?
5. *Dialogue.* What do the people actually say? What do the people say about each other?

Once you have determined as much of this information as possible, you should have a reasonable sense of the people involved in your scenario. Knowing the players can make a big difference when it comes time for you to analyze the situation.

The Situation

After you've sorted out the players in each scenario, including their personalities and relationships to each other, you must determine the situation at hand. The *situation* refers to the events that caused the ethical dilemma to occur in the first place. Figuring out the situation is not as simple as it might seem. You may quickly settle on what you perceive as the factors that caused the conflict in the scenario, but after more careful analysis, you notice key details that, when added to everything else, change the situation. You must carefully sort out the situation, much as you sorted out the players. List the following information, separate from the list of players, if it is provided:

1. *Upon what initial request or recommendation is the overall problem based?* For example, did an employer request a document? Is an employee trying to solve a problem?
2. *What series of events led to the request or recommendation?* List the events, briefly, in chronological order.
3. *What external factors might influence the course of events?* Perhaps a particular client is demanding attention or is related to someone in the

company and, therefore, expects special treatment. A co-worker's personal life might affect the situation, such as spiritual beliefs that color his or her perceptions.

4. *What internal factors might influence the course of events?* Consider the implications of the problem for other people in your company. It is possible that someone's promotion depends upon swift action or that the company may face a greater problem, such as an employee strike or a consumer lawsuit.

These and other similar questions should enable you to isolate the fundamental ethical situation that is occurring in a scenario. In addition, isolating the sequence of events leading up to the conflict of the scenario should help you to view the conflict in *context,* that is, seeing the conflict only in terms of the situation at hand.

The Conflict

After you have isolated both the players and the essential situation, you must turn your attention to the most important aspect of the ethical problem—the conflict. Determining the fundamental conflict in any situation means separating the key question—"Is this right or wrong?"—from all other aspects of the scenario. Doing so allows you to think about the question in a completely logical fashion *before* you add both the players and the situation back to that conflict. You cannot make a good decision about any ethical problem if you do not consider ultimately the conflict in context, in its workplace environment.

Isolating the conflict for close analysis, however, means really uncovering the question that is at the center of the ethical dilemma. This question, the *conflict query,* allows you to examine carefully and logically some hypothetical responses before you add to that question the players and the situation—two factors that always further complicate a difficult scenario. The nature of the conflict query varies depending on the scenario. To isolate the query, ask yourself these questions:

1. What is it about this situation that could be perceived as either *right* or *wrong*?
2. What action is someone being asked to perform that could challenge a personal belief or code?
3. What ramifications will a person face for carrying out the action that challenges a personal belief or code?
4. Will certain choices result in firing or promotion? What fundamental question causes a person to weigh those two opposite employment-related outcomes?

Determining the conflict in an ethical scenario is not unlike determining the conflict in a favorite story or book. Conflict means opposition, and opposition is at the heart of the ethical dilemma. Isolate for yourself the opposition, the question of right or wrong, and come to some personal conclusions about resolution before you move to the larger issues in any discussion of ethics.

EXPLICIT ASPECTS OF ETHICAL SITUATIONS

The explicit aspects of ethical scenarios really refer to those direct situations occurring in any professional setting that challenge your personal sense of ethics. While there are no easy answers in the following examples, you must examine your own ethical code to address them. Remember that it is not uncommon in business and industry to be asked to do something that you believe to be contrary to your code of ethics. When this occurs, you must be prepared to study the situation, to evaluate the variables, and to make a good decision based on careful consideration.

The following four examples of explicit or direct aspects of professional life should familiarize you with the kinds of situations that can arise (and likely will arise) once you are in the workplace. These direct aspects are also relevant to the scenarios at the end of this chapter. Are the people involved in the scenarios being asked directly to write or to do something that seems to challenge their personal code? Close examination of the explicit aspects of scenarios can help you to arrive at more thoughtful and carefully considered responses.

Direct Instructions

In some situations, employees are given direct instructions by people in positions of authority to perform an act or write a document that may challenge a personal ethical code or belief. Consider, for example, how you might react if your boss directed you to eavesdrop on a conversation a co-worker was having to determine whether or not the co-worker was using company time for personal calls. Consider, also how you might react if your boss asked you to draft a memo to the employees announcing future layoffs because of financial problems. In such situations, think about the following questions:

1. Are the direct instructions tied to a matter of life or death? Will your compliance with the direct instructions result in potential loss of life or loss of health?
2. Are the direct instructions tied to a personnel matter? Will your compliance with the direct instructions result in you being marked as an informant?

3. Are the direct instructions relatively trivial in nature? Will your compliance with the direct instructions result in a minor policy change or a shift in procedure? There are more scenarios than those listed, but these three will prepare you for the kind of direct instructions you might receive in the real world.

Direct Benefits

Unlike direct instructions, direct benefits refer to those situations in which your potentially unethical behavior will result in direct benefit to you. Usually employees are motivated to act when the company benefits because this is part of the job; promotions are often based on such activities. What happens, though, when the benefits are strictly personal? Consider, for example, a scenario in which you have an opportunity to sell company secrets to a competitor for a sizable amount of money. What would you do if a competing company, located in another city, offered you an increase in salary just days after you renewed your contract with a present employer?

You might find it easy to respond to both these situations ("Yeah, I'd do it if the price were right!"), but really evaluate what is happening and what your actions might mean to others. Ethical technical communicators must carefully consider situations of direct benefits because these can, in some scenarios, seem the most alluring and, therefore, the most problematic. Consider all aspects of the situation before making a decision.

Direct Competition

You and your friend Bob have been competing since college. Now, five years later, you are both up for the same promotion. You've been told by a company insider that she can virtually assure you the promotion if Bob's monthly financial report "happens" to get lost. You want this promotion. Will you help "lose" Bob's report? There are similarities between direct competition and direct benefits. In this scenario, you will directly benefit if you are promoted. The important distinction, though, involves a competitor, in this scenario, Bob.

In direct competition, you are often faced with ethical choices that involve co-workers, business competitors, even friends. You must evaluate the situation carefully to make sure that your response to competition is fair and equitable. Competition is the driving motivation behind the free enterprise system in the United States. Indeed, in the United States competition in the workplace can still result in incredible success for those who are hardworking and motivated (just ask Bill Gates, president of Microsoft). Responding to issues of ethical competition, however, can be tricky depending on the situation. Look, therefore, at all aspects of a situation to ensure that your response is based on ethical communication and good reasoning.

Direct Association

One of the most difficult ethical areas to sort through is the "group dynamic" associated with direct association. Often in business and industry, you will find yourself working in collaboration with others. Sometimes that collaborative effort may be brief (a few days or weeks), but sometimes the teamwork extends into several months or even years. When collaboration works effectively, people come to know and value each other. It is in this environment that issues of direct association can become problematic.

Imagine, for example, that you have been working with a group of colleagues for six months on a new child safety seat design. The design is revolutionary, but the group learns that a competitor is just about to unveil a product similar to yours. The group earnestly wants to go public with the design first, so the group leader suggests skipping the final phase of testing, which will speed up the process considerably. You are very concerned about this, but the other four members of the group prevail upon you to give in and go along with the group. Will you?

Direct association is problematic because it raises in us issues of loyalty and commitment to others. In some instances, groups become small families as they work together toward a common goal. Resisting the will of the group can be difficult, then, when ethical questions arise. It is your responsibility as an ethical thinker and communicator to evaluate any collaborative situation for hints of future problems. Carefully consider questions of loyalty to the group when you are being asked to do something that is completely contrary to your personal code.

TIP

As you evaluate the scenarios at the end of this chapter, look for instances of explicit situations like *direct instructions, direct benefits, direct competition,* or *direct association.* When you spot such aspects, carefully document them. Such information will be useful when you draft your Solution Defense.

IMPLICIT ASPECTS OF ETHICAL SITUATIONS

Just as the explicit aspects of ethical situations can be problematic, so too are the implicit aspects difficult to identify. The word *implicit* refers to those indirect or implied problems that can be, in many scenarios, hard to isolate. If the explicit aspects are direct and obvious in most ethical choices, the implicit aspects are less obvious but equally important. In each of the following four areas of implicit or indirect consideration, think carefully about how you might act in a similar scenario, and familiarize yourself with the kinds of situations that can arise once you are in a workplace environment. These four areas are also relevant to the scenarios at the end of this chapter. Are the people involved in the scenarios acting in a way that indirectly violates procedure or code? In each scenario, study the situation and evaluate the variables. Close examination of the implicit aspects of a scenario can help you to arrive at a more thoughtful and carefully considered decision.

Indirect Loyalty

Though many of us make decisions based on loyalty, there are situations in which we are not even aware that we are placing loyalty ahead of our sense of right or wrong. Let us imagine, for a moment, that you are a midlevel manager of an engine manufacturing company that has hired three generations of workers from your family. Your family ties to the company, therefore, are very strong. You are approached by a line worker who has, for some months now, been removing foreign plates from engine assemblies and replacing them with "Made in USA" plates to meet state requirements. The worker is beginning to feel guilty about the process and approaches you with his concerns. You had no idea. How will issues of loyalty play in your decision to follow up?

Loyalty is a difficult concept to define, but it is an important aspect of the workplace. Without some loyalty to a company or to an individual or group, most of us would have little interest in our work beyond the paycheck. The crucial issue here, though, is not loyalty itself (a positive workplace element), but indirect loyalty that causes you to make decisions based not on good ethical reasoning but on a sense of commitment that can cloud clear thinking.

Indirect Rationalization

The indirect rationalization is perhaps the most difficult of all indirect aspects of ethical choices to define because it is really the one to which most of us like to believe we could not possibly fall victim. In the indirect rationalization, you acknowledge the nature of an ethical dilemma and make an ethical choice that violates your personal code. You do this, though, because you rationalize the outcome. In this situation, you are able to fool yourself into believing that even though you are not doing the "right" thing, everything will turn out satisfactorily anyway.

Imagine that you are a technical writer for a wastewater treatment company. You know that the company has been pumping small quantities of untreated, bacteria-laden wastewater into a local stream. As you prepare the yearly report on the company's activities, your boss asks you not to mention the small quantities of untreated wastewater because they are so minuscule as to be insignificant. You are unsure, but you leave the material out of your report because you convince yourself that such small amounts surely couldn't have a harmful effect on the environment. In this scenario, you are rationalizing a "happy ending" to an ethical choice that may or may not turn out to be a reality. Carefully evaluate, therefore, your own motivations as you make ethical decisions.

Indirect Challenge

The indirect challenge, like the rationalization, involves your perceptions of a situation. If an indirect rationalization means you convince yourself of a positive outcome to an ethically difficult scenario, the indirect challenge deals with the ways in which you believe others might perceive you if you don't make the "right" decision. Whether we like to admit it or not, we all fall victim to peer pressure, and often we make poor choices as a result. In the workplace, poor choices can be the result of how you believe others will perceive you if don't go along with the crowd, make a "gutsy" decision, or otherwise demonstrate that you have the fortitude to make a sometimes unethical choice.

Consider the following scenario. You are the personnel manager of a small construction company that employs about forty workers, including clerical, midmanagement, and administrative people. Many of your employees smoke, though you do not. You decide, after consulting with the assistant personnel manager, to establish a smoke-free environment. You realize that this will likely be an unpopular decision, so you tell the assistant personnel manager to write up the reasons why in a memo for general distribution. She, a smoker herself, is not especially happy about writing a policy memo that really reflects your decision. You have the uneasy feeling that your authority may even be challenged because you asked someone else to draft the memo. Are you just being paranoid? Is this a good ethical decision? If it is, should you write the memo? Evaluate a situation like this very carefully to determine whether or not your own perceptions might be coloring a good decision-making process.

Indirect Nepotism

In indirect nepotism (favoritism to relatives), you make decisions not based necessarily on what is best, but instead based on what is most beneficial to a family member. It's not that many of us haven't gotten jobs through relatives. Probably many of you have enjoyed summer employment thanks to a relative, close friend, or even the friend of a friend. But what happens when that commitment to family clouds your judgment? Even though you may not be aware of what you are doing, making decisions that favor family or close friends can be ethically problematic.

Suppose you are, once again, the personnel manager of a small construction company. Because of your position, you know that the company will soon have an opening for a receptionist at the front desk—someone who can answer the phone, do light typing and filing, and keep track of appointments. You think immediately of your niece, Kelly, who has been struggling to get by since her husband abandoned her. Do you encourage her to submit her application to the company? You know that Kelly has minimal typing skills because all her work experience has been in waitressing. You, along with your small staff, review the applications and isolate three strong candidates and Kelly. You are very tempted, in your position, simply to "pull rank" and hire her right away. This is a tough ethical choice.

Deciding the correct answer to this dilemma, and others like it, is not easy. Thus, you, as an ethical communicator, must consider all the factors in any situation—either real or imagined in a scenario—before making a decision. Recognize the implicit or indirect factors that are also a major part of any ethical problem. Always evaluate each scenario or each problem carefully and logically before arriving at a decision.

TIP

As you evaluate the scenarios at the end of the chapter, look for instances of implicit or indirect situations, including *indirect loyalty, indirect rationalization, indirect challenge,* and *indirect nepotism.* When you spot such instances, carefully document them. Such information will prove useful when you write your Solution Defense.

DETERMINING THE RAMIFICATIONS

Perhaps the most important aspect of making a good ethical decision involves careful examination of the ramifications or results of those decisions. If you choose one solution over another in any ethical situation—real or imagined—consideration of the likely outcome must play a key role in the decision-making process because it is really the outcome that most of us care about. (You've heard the old adage about "the ends justifying the means.") But is this really the scenario? Do we carefully consider the ramifications when making ethical choices? Consider the following ethical dilemma and the ramifications that could result from either choice. You decide which choice makes the most sense, given all the factors we have discussed thus far, but most especially the ramifications of the scenario.

You and a close family friend are advertising executives for a New Jersey company specializing in ad campaigns for a Japanese firm selling electronics in America. You are not that wild about the product you're being asked to advertise—stereo speaker components—but you accept that your job means sometimes working on projects that are not personally satisfying. You and your ad executive friend have been sent several sets of the speakers to try at home.

Accidentally, though, a company representative from Japan sent an extra shipment of electronic equipment, including compact disk players, VCRs, and a camcorder to your friend. You are shocked when your friend reveals that he received the merchandise a little over two weeks ago. You ask him what he did with all the merchandise, assuming that he returned it to your boss, the vice president for marketing. Imagine your surprise when you discover that your friend gave away the merchandise to friends and family members. He implores you not to say anything. You wonder, though, if the home company won't want the merchandise returned. Your friend notes that he will simply deny ever receiving anything, so there's no way to trace it. "We'll just tell them that all that stuff got lost in the mail," he says. "Besides, the stuff is garbage anyway."

You are uncertain and feel as though you just got involved in a kind of conspiracy. You sit down calmly to try to reason through the ramifications of both choices facing you—saying nothing and telling everything. On the one hand, if you say nothing, you will not alienate your good friend and with any luck, nothing will happen. On the other hand, if you tell your boss everything, you can possibly intervene before your friend gets in a lot of trouble. If your boss does get involved, you can hopefully convince your friend to package up the merchandise and return it.

What would you do? There are potential ramifications in this scenario that extend far beyond the few noted above. Develop two separate lists of ramifications—one for silence and one for telling—and come to some ethical conclusions about how you would handle a scenario of this sort. Remember that evaluating the ramifications of a scenario can be very useful as you try to make a reasonable, logical choice.

CONSIDERING THE FOUNDATIONAL QUESTIONS

Consider the following foundational questions before you make any ethical choices for the scenarios at the end of this chapter or real workplace problems. These questions are very important as you evaluate individual situations and attempt to come up with a solution that is hopefully best for everyone concerned.

Questions about the Situation

1. Is human life at risk in this scenario? Could anyone be seriously injured because of your decision?
2. Is the environment at risk in this scenario? Will your decision cause harm to the local or national environment?
3. Is the law being violated? Could your decision have legal ramifications for you or your company?
4. What is the impact on society? Will your decision cause change great enough to help or harm a portion of society?
5. What is the impact on future generations? Will your decision leave behind problems that the next generation of decision-makers must face?

Questions for the Writer

1. Are you being honest? Are your actions based on honesty or on other factors (money, job, security, loyalty, fear)?
2. Are you violating a personal code of ethics? Will your decision bother you later because it is contrary to your beliefs?
3. Will your actions hurt anyone outside the company? In making your decision, will someone at another company lose a job, a promotion, or a client?
4. Will your actions hurt anyone inside the company? In making your decision, will someone within your company lose a job or a promotion?
5. Can you live with your decision? Will your decision hurt your work performance?

These and other foundational questions are germaine to the process of making good, ethical decisions. What follows are a series of scenarios devised to challenge your best instincts. Consider the factors covered in this chapter as you generate solutions for these scenarios, and remember: solving ethical dilemmas is among the difficult of tasks you will face in the workplace.

Exercises for Classroom Discussion

1. You are a technical writer for the automotive industry. You have been given the task of writing a form letter that will be mailed to all owners of the Markham V6, a popular sports utility vehicle. The vehicle was recently tested by a nonprofit consumer's group, which released findings that suggest the Markham V6 has a tendency to tip over when the operator makes sharp turns at about 40 miles per hour (mph). In the letter you must reassure owners that the car is safe and does not demonstrate any tendency to tip onto two wheels unless the operator is making a very sharp, abrupt turn in excess of 80 mph. As you begin to draft the letter, you decide to review engineers' and testers' notes on the situation. To your surprise, you find handwritten notes indicating that tipping is in fact present at speeds as low as 50 mph when the sharp turn is executed. Sort through this brief scenario, decide upon a solution, and be prepared to justify your response.

2. You are a technical writer at A & M Construction. Your boss, the Personnel Manager, tells you that he needs some help drafting a document that will be given to all employees on the problem of excessive photocopying. Employees currently make copies and "log in" the number of copies on a handwritten sign-in sheet. While some employees faithfully log in copies, others appear to make large numbers of what your boss suspects are personal copies. The yearly budget for copying is rapidly approaching zero, and your boss is ready to take a hard line with the employees—pay a nickel for each personal copy or the company will reserve the right to dismiss employees caught making excessive personal photocopies. You really dislike writing such a threatening letter. Analyze the ramifications of the policy and be prepared to defend your solution.

3. You are the new assistant personnel manager in a small computer company. You like your boss, most of your colleagues, and your assigned responsibilities; you like working with people. Your boss, the personnel manager, approaches you one day to discuss the possibility of a new security system for the company. You are quite surprised to learn that she wants to install hidden cameras on the showroom floor to videotape both customers and employees. In fact, your boss suspects that employees are the ones stealing computer components. The thefts are costing the company thousands of dollars, so she is understandably upset. You understand the problem, but you are very concerned about the workplace environment such a step would create. Your boss is puzzled by your hesitancy. Workers on the showroom floor are wary and suspicious when they hear "through the grapevine" about the possibility of cameras. Study the ramifications of this scenario carefully and be prepared to defend your solution.

4. You have been hired as a public relations writer for an ad company. In addition to preparing graphic designs and layout for ads, you must draft the written portions of those ads and the letter that will accompany mass-mailed flyers. You are a little unsure of drafting the material for one product you know well—the TF-111 String Trimmer. You purchased one of the rechargeable, battery-operated string trimmers a year before. As you reviewed the notes from engineers and the testing lab manager, you were surprised to see the trimmer described as completely safe to use around trees because of the lightweight polymer string. In your own experience, the string had in fact damaged young trees with either thin or new bark. Your eight-year-old silver maple was seriously damaged when you accidentally bumped the tree trunk with the trimmer. You are to write ad copy for the product describing it as non-damaging and even "gentle" to trees. Carefully examine the situation in this scenario and the ramifications and be prepared to defend your solution.

5. You are the police chief of a small southern police department. The building that you and your five deputies, a secretary, and the assistant chief occupy is small but has served you well for the fifteen years you have been chief. The mayor of your small town, who faces reelection in one year, has decided to seek funding for a new police building because, as he has told you in conversation, he believes that a new building would be a good, high-profile project for the community. Besides, he dislikes the old building and wants it demolished. He has asked you to do a write-up of the department's needs so that he can approach the head of the state legislature's appropriation committee, his good friend. You are quite satisfied with the present station, but you also begin working on the report for the mayor. Several days later, and to your great surprise, a citizen's group storms into the police station demanding to know why a building that the townspeople want on the National Historic Register is to be torn down in favor of a new police station. All of a sudden, you are in the middle. Study the scenario carefully and devise a solution to this situation. Be prepared to defend your solution.

SCENARIOS FOR ETHICAL CONSIDERATIONS

The Carpet Treatment Problem

CONSIDERATIONS

In this scenario, you are a technical writer working on literature for your company's new product. This puts you in the interesting position as the person who is the last to look at all the documentation before the product is manufactured and sold. When reading this scenario, keep the following questions in mind:

- What is the nature of the information you are gathering—reliable, suspect, ambiguous, subject to interpretation?

- What do you do with the information provided by someone who does not want to be identified?
- What effects (on the company, its management, the researchers, and your own position) could your actions have?

You are the head technical writer for New Age Chemical, which employs over one thousand people in the small town of Orion, Pennsylvania. New Age recently downsized to increase its efficiency and profits. The top brass cut everyone's staff (including yours), eliminating many middle-management positions as New Age competes with chemical manufacturers who moved to third-world countries and are undercutting New Age's prices.

Research and Development (R&D) at New Age invented a new product that might help the company's situation. They created for any new or in-use carpet a treatment that provides virtual stain-proofing and greatly reduces carpet wear. The company is touting the chemical treatment, known in New Age's lab as Formula 14, as the best available, and they expect it to exceed a rival company's antistain technology.

You are overseeing the creation of all the product information for the carpet treatment. This includes instructions for using the product, and the "sell sheets," or literature that tries to convince customers of the carpet treatment's benefits and unique performance. You follow normal procedure and collect all the documents from R&D on the carpet treatment project, such as laboratory notebooks, internal memos, and the paperwork from the Occupational Safety and Health Administration (OSHA), a government agency that authorized New Age to manufacture the carpet treatment.

During your work on the product information, you notice one particularly interesting entry in the laboratory notebooks. It reads:

November 9: tested Formula 14 on sample 23B. Substance precipitated. Very fine smokelike gas noted up to one-half inch from sample. Stopped when formula dried—need more tests on B lot #22–26. Determine gas composition. Removed precipitate. Posttreatment: sample 23B tested within Formula 14 performance parameters. RFH.

You look through the lab notes, trying to figure out what this rather vague entry means. You discover that the lab tested carpet samples in two groups: lot A was untreated carpet made of different materials (wool, polyester, various blends, etc.), while lot B consisted of carpets treated with various chemicals, from stain resistors to cleaning products. Looking at the

November 9 entries, you see that the researcher requested tests on five samples to "determine gas composition" while only noting gaseous emission from sample 23B. You find neither evidence that B samples 22, 24, 25, or 26 were ever tested nor was the precipitate (solid particles) tested.

You call Ron Hayworth, the lab researcher whose initials appear at the end of the November 9 entry.

"Lab, Beth speaking," a voice answers.

"Ron, please," you say.

"I'm sorry," Beth says. "Ron is out sick today. May I leave a message?"

"No," you reply. "Could I speak to someone familiar with lot B testing on the Formula 14 project?"

"Sure," Beth replies. "Bob and Joanie aren't here today, but Kevin, the technician, is available. One moment."

You wonder why most of the Formula 14 team isn't around, especially because the company top brass recently pushed the production deadline up a month. *Why aren't they working overtime down there?* you think. After a few seconds, you hear the tell-tale click of someone picking up the phone. It's Kevin, the technician. You quickly explain your reason for calling, read him the November 9 entry, and politely ask for an explanation.

"Oh, yeah," Kevin answers. "We grouped each lot into subgroups of similar composition, and tested one sample from each subgroup. If there were any problems within a subgroup, we tested each sample."

"Why weren't tests on 22 through 26 done, then?" you ask.

"I'm sure they were," Kevin replied. "We had so much to test, we stopped noting results in the reports unless there was something unusual. We also changed a few things about Formula 14 in mid-December, so that problem's probably gone now."

"You're supposed to note *every* test result," you remind Kevin.

"Well," he says a little sheepishly. "Things have been a little, well, rushed down here."

"I'm sure they are," you say. "When will the regular researchers be back in case I have more questions?"

"I don't know," Kevin answers. "Ron, Bob, and Joanie are out sick today. We seem to have some kind of virus running around down here."

"Yuck," you say. "I won't be visiting, then."

"You wouldn't want to," Kevin laughs. "We're working on making Formula 14 smell better. We didn't realize it stank until we started working with it outside the containment vats. My wife makes me take off my work clothes the minute I get home, and then she stuffs them in a garbage bag and takes them to the laundromat. She won't even wash any of it at home!"

"Hope you work it out, soon," you say, glad for the first time that you are a writer, not a researcher.

After hanging up the phone, you look again at the laboratory notebooks. As Kevin said, entries past mid-November taper off, an indication that the researchers stopped noting every test result. You also notice lengthy absences from several of the people working on improving the formula's smell.

You find that odd and call Marty Thomas, head of New Age's human resources department. You explain your situation to her and request some help, specifically why the R&D workers called in sick.

"Bronchial irritation is what the company doctor came up with," Marty says."

How serious was it?"

"Not bad," Marty says. "It went away after a couple of days."

"Thanks Marty," you say, preparing to hang up.

"Wait a minute!" she says. "You might want to know that one of the technicians asked for a transfer from that project. He said that something in the carpet treatment was irritating his lungs, and he wanted off that team."

"Really?" you say. "Did anyone take this to management?"

"No," Marty answers. "You know how important this stuff is to the company. And you didn't hear about that transfer from me, OK? You say you heard it on the grapevine."

"Got it," you reply. "Thanks."

You realize that something might be wrong here. From experience, you know that health problems occurring during product development often show up in the people who use the product. You wonder about people being around Formula 14 when it's still wet—maybe you can put a warning in the product literature. You decide that you have no choice but to take your concerns to Bill Quayle, the Vice President of Operations, who is your immediate supervisor.

You manage to squeeze into Bill's meeting schedule, and walk into his office feeling a little illprepared. "How are you?" he asks, inviting you to sit in one of the overstuffed chairs across from his desk. He sinks into his chair, looking so relaxed you think he might put his feet up on the desk at any minute.

Good, you think. *He's in a receptive mood.*

"I'm fine," you answer and decide to get right down to it. "I'm worried about a possible health hazard with the new carpet treatment product."

You show him the laboratory notebooks, particularly the November 9 entry, and point out the absence of the researcher due to bronchial irritation. You also mention that you heard about one technician who wanted off the project because he had lung problems.

"Do these notebooks say anything else about possible hazards?" asks Bill.

"No," you answer, "I spoke to a lab technician who said they stopped making notations on results unless there was something unusual. But that means there's no evidence the tests requested in the November 9 entry were ever done, or that we know that the precipitate and the gas are harmless, or that they don't occur now with changes in the formula."

"What do you suggest we do about this?" Bill asks.

"I think we should find out if there are adverse reactions to some pre-treated carpets before we ship any of the product," you answer. "We don't know if there are long-term effects on people in houses with this carpet treatment or on the manufacturers using it."

"You know we're on a tight deadline," Bill says.

"Yes."

"You know running these tests you're talking about could take months."

You don't say anything but continue direct eye contact.

"OSHA already approved our application to manufacture and sell the product," Bill continues.

"What would the financial impact be if this stuff causes health problems?" you ask. "The manufacturers and customers could sue. Our employees could sue if manufacturing the carpet treatment is harmful."

"It's a risk I think we will have to take," Bill answers. "Our profits for this quarter, our reputation, our ability to keep paying our employees—everything depends on this carpet treatment going to market in exactly two weeks. We've already taken in thousands of dollars in advance orders from distributors."

Bill taps his fingers to his chin, obviously making a decision. "Just get the product information done, and we'll start health tests *while* we manufacture. I'm sure it's nothing serious."

"Should I include a warning about being around the product while it's drying?" you ask.

"Did R&D call for that?" he says.

"There's nothing specifically mentioned, but it wouldn't hurt," you say.

"But it could tell buyers that there's a problem," Bill says. "I tell you what. Why don't you leave it out for now, and we'll discuss it when you bring me the final draft."

"Sure," you say.

"I expect you'll have that done within the next few days?" he asks.

"Yes," you say, gathering your papers.

"I'm glad you brought this to me," Bill says, smiling.

You leave his office, knowing that you must do something more even if you don't put a warning in the literature. What about the company employees who work with the stuff? *Won't they have the same problems as the R&D technicians?* you think.

You think over your options. Bill's immediate supervisor is none other than Carl Pacer, the Chief Operating Officer of New Age. You could write to Mr. Pacer, but that would mean going over Bill's head, which you know he wouldn't like. *Whatever I do,* you think, *I've got to start documenting my concerns.*

Suggested Next Steps

- Use the guidelines for handling ethical dilemmas included in this chapter to identify all the issues and concerns involved.
- Decide upon a course of action, and consider all the possible effects your action may produce.
- Write to *someone* (you choose who you feel is most appropriate for the course of action upon which you've decided) to document the problem and your reaction to it.

Environmentally Friendly Frost Spacers

Considerations

In this situation you are indirectly responsible for the safety of every car driving on a particular stretch of road. When reading this scenario, keep in mind the following questions:

- What are the political and financial relationships between your department, Madeline's office, the federal government, and Amex Corporation?
- How reliable is the information you receive on the frost spacers?
- How free are you to use all the pieces of information?

"Hey," Bob yells across the room. "Did I tell you that we had another complaint on that spacer out on Highway 67? Looks like this time the guy's gonna try to sue the state for repairs to his car. I guess he broke a tie-rod when he hit that concrete ridge that's developed there by Rigby's farm. We gotta get a crew out there soon. Oh, by the way, welcome back from vacation, boss."

"Yeah, thanks," you reply sighing. "Let me check my e-mail and answer some messages, and we'll talk." Phoenix and the warm desert sun seems like a distant memory now. April in the Upper Peninsula, you think to yourself. Everything's breaking up, including the roadways. "I have to remember to vacation in April next year," you mutter to yourself under your breath.

You know only too well of the problems the Upper Peninsula of Michigan has been experiencing with frost heave during the last two extremely rough winters. As Project Coordinator for the state of Michigan—U. P. region, you've dealt with eroding roadways and crumbling interstate highways for years now. Yet the last two years posed more problems than you imagined in your worst dreams. The frigid temperatures, above-average snowfall and horrendous winds have really taken a toll on the already weathered roads and bridges. Now some guy wants to file suit against the state for a broken tie-rod? *Fine,* you think.

Later in the day, Bob fills you in on the spacer situation. Last year, Bob believed the spacers that absorb stress and reduce frost heave looked pretty good. The spacers held up for the most part—your department had to replace only a couple in heavily trafficked areas—and the crumbling was virtually nil. This year, though, the shifting temperatures caused big ridges to develop around the seams of the spacers. In a couple of spots there are concrete heaves around the seams five or six inches high.

You are not surprised. You remember problems with highway spacers from when you worked for the state of Alaska. Usually the ridges that develop around the seams are no big deal, but once in a while, when the weather's severe enough, the seams just keep growing. The stress to vehicles is serious enough to cause damage and sometimes serious accidents.

You decide to tackle the spacer problem right away, and not just because travelers might start suing your department. Madeline, your state government contact, has offered money that might, in part, solve the problem with spacers. The state of Michigan recently contracted with Amex Incorporated of Grand Rapids. Amex has developed a new material that replaces the conventional concrete–asphalt mixes that go into the making of those spacers. This new material is a mixture of concrete and recycled plastics.

Madeline is pretty enthusiastic about this new process. She has told you that spacers are environmentally friendly and she was able to get federal grant money to help secure the project funds. After your last conversation, Madeline asked you to have your road crews ready to go in May to start pulling up the old spacers and replacing them with the new material. You think she wants your area to be a test case for other cold-weather cities.

You scroll back through your e-mail messages to reread the one from Madeline. *Boy,* you think, *this is really a big deal for her—not to mention Amex. That grant money is really going to be a windfall for the company. If this new stuff works,* you think, *everyone will come out ahead.* Madeline wanted you to reply to her within a week, and when you go back into your e-mail to write down her address, you notice a new message that's arrived since your meeting with Bob. It's an e-mail from your old friend Tim Henderson, an engineer with the Alaska State Highway System. You worked with Tim during your stint in Alaska and came to respect his judgement and advice. You decide, after some thought, to get some feedback from Tim on the Amex spacer mixture, but you want to talk to him. E-mail is great, you think, but for you nothing replaces the immediacy of the telephone. You call Tim, and to your surprise, you catch him.

"Hey, how's it going?" you ask as you hear Tim's surprised voice. "How's life treating you?"

"Oh, I can't complain," Tim replies. "Well, actually, I can complain, but I won't."

The two of you laugh over old times, and you get to the point.

"Tim, we just got some grant money to try an experiment here, and I'd like to run it by you for a little feedback." You explain to him about the Amex product, the grant, the environmentally friendly aspects of the experiment, and Madeline's enthusiasm to try something new.

"Well," Tim replies hesitantly, "I wish I could congratulate you and Madeline on this, but frankly what you've just described is not all that new. The Department of Alaska Highways and Transportation in Anchorage tried a very similar product roughly five years ago with mixed results. The spacers definitely didn't shift and heave as much in weather extremes, but they did tend to break up more quickly than conventional concrete spacers. What we ended up with were more and larger pot holes."

"Are you sure the product is essentially the same, Tim?" you ask. "Are you sure we're talking about a concrete–asphalt plus recycled plastics mixture?"

"Yep, we're talking about the same product. I can't remember the name of the company we worked with, but it sounds like the same stuff."

"Well," you respond, "maybe the percentage of plastic to concrete is different."

"Maybe," Tim adds, "but that doesn't necessarily make this stuff new or improved. Besides, the mixture can't be too different from what we used here in Alaska. Too much plastic and you're going to have problems."

"Thanks, Tim," you say in closing. "Do you maybe have any other information you could send me on this? Madeline needs to know about this."

"I'll see what I can find," Tim responds. "I'm not going to be able to send much, though, because the whole thing turned out to be a mess for the department. I'll be in touch."

You trade good-byes and hang up the phone. You weren't prepared for Tim's response.

A few days later, you receive the following letter from Tim:

I really can't send you anything official, but I can briefly relate what happened here in Anchorage when we tried the "new" substance you described on the phone. In 1990, the Department of Alaska Highways and Transportation tried a materials experiment to deal with the problem of frost heave in the region. The mixture we used, a formulation developed at the University of Alaska and marketed through a company I'd prefer not to name, was 80/20 concrete to recycled plastic.

continued

The results of the mixture were inclusive at first. We had no real problem in the first year, but that was a mild winter. By the winter of 1991, though, we experienced not only heavy snow in the Anchorage area but extremely cold temperatures in February. By spring breakup, we noticed serious deterioration of the spacers, including potholes and crumbling around the edges of the spacers.

You know, I'm really not at liberty to send you specifics. The department absorbed some bad publicity over the situation. The daily paper, the *Anchorage Northstar*, did an article about the cost of the experiment to the taxpayers. I'm still not sure how the guy who wrote the story got the information, but we really faced some serious damage control after it came out. My boss wouldn't be happy about me sending this letter, so use the information to help make a decision, but don't say where you got it.

Hope all is well on your end. The winter's been pretty quiet here. Take care.

—*Tim*

After reading the letter, you wonder why Tim doesn't want to be quoted. You really need to share this information with Madeline, but you can't tell her where you got it. On the other hand, you could just ignore what you know and go ahead with the grant-funded project. More importantly, you have to write Madeline with some kind of answer, and soon.

Suggested Next Steps

- Review and work through the procedures for identifying the aspects of ethical dilemmas as outlined in this chapter.
- Based on the information provided, decide who and what you wish to believe (or not believe) concerning the experimental frost spacers.
- Decide how to characterize and justify your beliefs. If you are basing your beliefs on information you've been asked not to repeat, how do you use this information given in confidence?
- Write to Madeline in answer to her request for a "yes or no" on the project. Be sure to include any information you feel is relevant and appropriate for this situation.

The Heated Sidewalk Problem

Considerations

As a technical writer, you are often the person who writes the first documents for "general consumption," or outside audiences. In this situation, you have reservations about how your company is using a product in a particular situation. While reading, keep the following questions in mind:

- What are the major concerns of your company and of the college purchasing your product?
- As a new employee, what are the potential impacts on your ability to work with your colleagues on future projects?
- What information are you provided, and how reliable is that information?

This is a pretty big day for you. You've been employed as a technical writer for just six months at Michaels and Greenwall Associates, an engineering and architectural firm located in Madison, Wisconsin. You have enjoyed the varying writing jobs you've tackled, including reports, proposals, one grant, and even some public relations material. Today, for the first time, you meet with a large group of company engineers to begin gathering information for marketing a new product.

You stop by your office, drop off your coat and briefcase, gather your notepad and a few pencils, and walk down to the conference room for your 9:00 a.m. meeting with the engineers.

"Welcome. Sit down," says the lead development engineer Robert Cruzner as you walk into the room. You settle in, take out your materials, and listen to the conversation.

"We've invited you to our meeting today," Robert begins, "because we need a marketing letter written on our low-wattage electrical sidewalk system known as *Hot Blocks.* We're pretty excited about the product, especially for cold weather cities, and now that our testing phase is just about completed, we need the letter written."

"Okay," you respond. "I think I'll just absorb some of the conversation, and then ask a few questions. Will I have access to your notes and lab reports?"

"We'll give you copies of whatever we think you need to write the letter," Robert quickly responds. You detect a slight edge in his voice but decide you are being a little suspicious.

Robert addresses the group, "We've completed the preliminary testing phase on *Hot Blocks,* and we're pretty confident that this product is ready to go. In fact, the administration at Eastern Wisconsin University is ready to sign a pretty lucrative contract with us to install the sidewalks throughout the university." Robert looks at you, "That's why we need you to write up the letter about the product's features and benefits so quickly."

"Why a regional university and not a major university?" another engineer, Luther Blackwell, asks. "Usually the big schools have the money for something like this."

"Well, apparently a student was badly injured at Eastern last year," Cruzner continues. "She slipped on the ice as she entered the university's administrative center, fracturing her hipbone and breaking her hand. There's been talk of a lawsuit. I guess Eastern feels that it's demonstrating goodwill to install a system like this—before the case is filed."

"Are they ready to commit the finances?" Janice Blake, a systems engineer interjects.

"Not until they receive our report indicating that the product is safe, reliable, and ready to go," Cruzner adds. "I think we're at that point, but I want to go over the testing with you one more time."

For the duration of the meeting—about an hour and a half—Robert Cruzner evaluates all aspects of testing the *Hot Blocks* product. You take notes, writing down key figures and adding questions you want to ask later. You notice that most of the testing on the product has been in lab-approximated temperatures of freezing or slightly below zero. Nothing under 10 degrees below zero is mentioned. You need to ask Cruzner about that later. The meeting breaks up just before lunchtime, leaving you time to ask Cruzner a few questions before you retrieve your tuna sandwich out of the lunchroom.

"Robert, I have a couple of quick questions," you say as he's about to leave the room. "Can you stay just a moment or two longer?"

"Sure," he replies. "Let's do it now because I have meetings on this all afternoon."

"Okay," you say, "this will be quick." "The *Hot Blocks* product is manufactured here on the premises, right?"

"Yes."

"And the product is a premade concrete plate with low wattage circuitry running underneath the plate that then mates with the existing sidewalk?"

"Well, yes. It's a little more complicated than that, but for your purposes that's adequate."

You detect a little condescension in his voice, but you ignore it. "In the notes I took this morning, I only heard mention of the lowest temperature testing being something like ten below. Is that right?"

Cruzner pauses before speaking. "That's low enough for our purposes and the contract with Eastern," he says. "Write the marketing materials based on the comments this morning and the notes I will give you. You realize, of course, that I reserve the right to read your draft before it goes to Eastern."

Actually, you weren't aware of that. Your immediate supervisor is Jason Monroe, the Products Manager. To your knowledge you reported to no one else. "I'll give the marketing letter to Jason first, Robert," you reply. "If you want to review it, he'll have it."

Cruzner pauses and looks at you. The beginnings of a smile develop around the corners of his mouth. "That will be fine," he finally concludes. He turns and walks away.

You leave the conference room and head back to your office to begin logging your comments into the computer. You read through the materials that Cruzner and his team of engineers provided you, make notes on potential marketing aspects of the new product, and arrange your ideas in a rough outline. You look forward to drafting this marketing letter, the first of your professional career.

After lunch, you return to your office to begin drafting the report. On the top of your desk is an unmarked, unsealed envelope. You turn over the envelope and pull at the flap tucked inside the envelope. Inside you find two sheets of paper, photocopies of something, and nothing else—no note, no indication of who sent it, nothing. Your first temptation is to go straight to Jason, but you decide to settle down and read whatever's on the two pages before making any snap judgments.

The first sheet is a photocopy of a memo from Luther Blackwell to Robert Cruzner. You are surprised to see that it is dated February 18, 1996 —a month ago. You read the memo.

MEMORANDUM

TO: Robert Cruzner

FROM: Luther Blackwell

DATE: February 18, 1996

SUBJECT: *Hot Blocks*

Bob, I've had a chance to do that testing you asked me to do, so I'm attaching to this memo my notes. I tested the *Hot Blocks* product at temperatures ranging from 32 degrees Fahrenheit all the way down to 50 degrees below zero (Fahrenheit). You wanted me to test the run-off of melting snow according to temperature, and I see no problems in rapid runoff or accelerated melting. The overall system looks good.

At 50 below, complete system integrity cannot be guaranteed, but such temps are rare and pose no problem at this time.

Give me a call if you have further questions.

The second sheet of paper, stapled to the memo, is a photocopy of what you presume to be Luther Blackwell's notes:

Reports of low wattage circuitry and heated sidewalks (known as "Hot Blocks")
Luther Blackwell, Test Engineer

I like the durability of our product @ the freezing point. Snow of approx. 6" melts @ rate of 1/2" per hour w/ no discernible puddling of runoff due to rapid melting. Looks good to me.

Rate distribution chart below—
6" (32° F) 1/2" runoff per hour
6" (20) 1/3" rph
6" (0) 1/4" rph
6" (–20) 1/8" rph
6" (–30) 1/16" rph
6" (–50) and below less than 1/16" runoff per hour, but rare conditions.
(Few hairlines @ circuitry base, but minor and temps rare. Can't guarantee won't be cracks past a certain point.)

You finish reading the memo and the notes and you are amazed first of all that someone unknown to you "happened" to leave them on your desk. You are also amazed to discover that in fact temperature testing had been done at temperatures well below minus ten and with mixed results. You are well aware of the weather in that region and a fifty-below day is not out of the question. The new material raises many new questions as you prepare to draft this letter that is supposed to sell Eastern on the product. You have to think through carefully your response to this problem.

Suggested Next Steps:

- Compile all the information you have on the *Hot Blocks.* Decide how you want to characterize this information if you decide to write the letter to Eastern, and consider all the effects your letter could produce.
- Carefully consider your options. Do you write the Eastern letter, or do you produce some kind of internal correspondence that addresses your concerns (if you have any)?
- Think through the effects of each possible decision you could make, and create the document(s) you feel are appropriate.

The Mysterious Syringe

Considerations

Nurses perform dozens of procedures daily, often in hectic and complex situations, and sometimes mistakes happen. In this situation, you are asked to write a report that potentially will harm a co-worker. As you read this scenario, keep the following questions in mind:

- What are the social and political relationships in this situation?
- How reliable is the information you've received?
- Are you satisfied that you know what events led up to the accident described? If not, what would you do to gather enough information to satisfy you?

"Boy, I'm glad I got that apartment when I did," you mumble to yourself as you sit in traffic on a Monday morning. "If I had taken the place on Hill Street, I'd be sitting here all day." You sit, arm resting on the steering wheel, sipping a cup of coffee. You keep talking to yourself. You try to wake up.

When you finally arrive at St. Monica's Hospital, a midsized Catholic hospital located on the outskirts of Seattle's east side, you are still glad you took the intensive care nursing position three months ago. Even though you miss your family back in Minnesota, you wanted a change of scenery and an opportunity to put your medical training to good use. You were also fortunate to get the day shift, so it seemed like the perfect opportunity.

You get off the elevator on the third floor of the hospital, pushing through the familiar red steel doors that read "QUIET" and "INTENSIVE CARE UNIT." You walk over to the nurses' station, sign in for the day, and say hello to your supervisor, Beth, a woman in her midfifties. After listening to her update on the new patients and their status, you turn to leave the station and start your rounds.

"Wait. Before you go this morning, I need to talk to you about the quarterly nurses' report."

You pause. "I'm sorry, Beth, but I don't know what you're talking about. Was I supposed to read a quarterly report?"

"No, no," Beth replies chuckling. "Obviously nobody told you, but today's your lucky day. You get to *write* the quarterly nurses' report."

You can't believe you heard her correctly. You've only been on the job three months. How can you, of all people, write a report like this? "I don't understand," you reply with hesitancy. "Nobody said a word about it to me."

"Oh, it's no big deal, really," Beth says. "We have kind of an unwritten policy here that the new personnel get stuck with the job. When we don't have any new personnel, we rotate every three months."

"What is this report about?" you ask. Nobody said anything about writing reports when you decided to become an RN, you think to yourself.

"Well, the report presents an overview of the number of and type of patients we've had on the floor during a four-month period. In addition, the report includes an assessment of activities on the floor, including details of drugs dispensed, the general distribution of nursing duties, and any mishaps on the floor. I know it seems like a lot, but really it's not so bad because everything is already written down somewhere. Besides, we'll all help you with it."

"When is this due?" you ask with concern. Your workload is already plenty to keep you busy.

"It's always due by the end of the first week of the new quarter. That would mean that this four-month quarterly report is due by, say, the end of the next week. Don't worry, it's not that awful. We'll start on it tomorrow."

You sigh, accept the situation and perform your duties for the day. You aren't really looking forward to tomorrow.

After a night's sleep and a little time to think about it, you convince yourself that the chore won't be so bad. It will be pretty routine, you convince yourself, and besides, Beth will help. You arrive the next morning ready to start work on the project. Beth sets you up in her office, at her computer, and gives you the raw data for the report. "See," she says pointing to the spreadsheet on the screen. "All the patient-related data is already on a spreadsheet, so all you really have to do is generalize about the type of patient we've seen and whether the numbers have gone up or down, stuff like that. Even the drug dispensing information is on spreadsheet. Just check to see that everything looks standard. Really, the only tough part is reading through the notes nurses and doctors have left on small mishaps, little accidents, and the like. We only have one mishap that will result in a citation for a nurse, so the task isn't too bad.

You smile weakly at her as she leaves and closes the door behind her. Somebody will be cited as a result of your quarterly report? Great, you think. Might as well get started. You write a general introduction for the quarterly report patterned after one another nurse, Mary Betts, wrote during the last quarter. By lunchtime, you have looked at the spreadsheet material and written up a fairly standard review of the period. Everything looks normal. You plan to start on the accident notes after lunch.

You return from lunch recharged. You finished the bulk of the spreadsheet material, which left only the accident reports. How bad could they be?

You close the door to Beth's office and get started on the routine mishaps—accidents in logging in medications (all corrected and initialed), one mix-up in medical reports when, coincidentally, two Robert Jones were on the ward at the same time (problem corrected), one janitorial accident involving spilled cleanser in the hallway (no one injured), and finally one report of a dropped, disposable syringe that was lost but later found after a patient stepped on it and got a needle stick. The patient, as a result, had to go through a battery of tests for AIDS, hepatitis, and other bloodborne diseases. You felt a little uncomfortable on that last incident because your report had to cite the only nurse who had really made an effort to befriend you, Barbara Price, a late night nurse whose shift overlaps yours. The notes cited Barbara as the one responsible for not properly disposing the syringe. You finish typing up the report, concerned about what the mishap report might mean for Barbara, and then stop.

The syringe case rattles something in your memory. You begin to remember Barbara telling you about an incident that occurred shortly after you arrived. The doctor on the shift, Marjorie Peters, administered an injection of antibiotics to a patient injured in a construction accident. When the patient began gasping for breath, at least as you remember the story, Dr. Peters placed the syringe on the sink top near the patient's bed and began to deal with his breathing problem. She apparently forgot the syringe in the ensuing commotion, and it dropped to the floor undetected where it remained for several hours until the patient unexpectedly got out of bed. When the patient stepped on the needle and yelled for help, Dr. Lyons (another shift physician) noticed it, picked it up, and reported the matter to Beth. Beth logged in the mishap as Barbara's mistake, since Barbara had been the last nurse to check on the patient before the incident.

You really don't want to shake things up because you are new, but you are almost positive that Barbara wasn't responsible for the problem. Should you go to Beth with this? Maybe, you think, you should just finish the report, turn it in, and not worry about it. What about Barbara, though? You decide to look at past reports to see if there's some other way to handle this problem, and you find that, in some cases, the report writer attached a memo containing additional information related to accident incidents. In each case, though, the writer was Beth, the highest ranking nurse in the ward. You decide to take the high road and start documenting your problem by writing to Beth, explaining the situation and the possible ramifications if you turn in the report with a potential error of this nature.

Suggested Next Steps

- Examine all the information you've received on the syringe incident. Decide what you will (or will not) believe concerning who should be held responsible for the incident.
- Carefully consider all the possible effects of your actions.
- Decide what you will write to Beth, based on your beliefs and any other considerations you may have.

Soil Incineration: The Protestors

Considerations

In this portion of your work on the incineration project, you experience firsthand the fear and concern of the people who will live near the incinerator. While reading this scenario, keep the following questions in mind:

- From the information provided in the previous chapter, what is the potential impact of the Protect Our Land (POL) organization on the project?
- Why do you think Jack MacGregor is so upset with your interest in the environmental aspects of the project?
- What are the implications to the project, your company, your job, and your ability to work with your colleagues on other projects should you pursue researching environmental issues?

It's been a few days since you last wrote to Jack MacGregor about the soil incineration project. You've been working on other documents for Donnelly Engineering in the meantime. You decide to go out for lunch, and you soon are standing at the counter of the Chicago Deli, a favorite lunch spot for downtown workers. After ordering a Swiss cheese and ham on homemade rye bread, you sit down at a corner table and begin to work on the huge sandwich.

"Is this seat taken?" a gruff voice asks. You look up and see a tall, stocky man dressed in jeans and a flannel shirt. You guess his age to be about fifty, but it could be less—his face and large hands indicate he worked outside most of the time.

"No," you answer, and he sits down.

"Carl Maki," he says, holding out his hand. You introduce yourself during the handshake.

"You work for Donnelly, I understand," Carl says.

"Yes. How did you know that?"

"Oh, we have a list of everyone working on that pollution project," Carl answers, giving you a hard stare.

"We?" you ask. "You must be from that organization, Protect Our Land."

"Sure am," Carl says proudly. "POL is the only thing between me and losing my dairy farm. I live on the Shewanee River, downstream from where you guys will be putting in a toxic waste dump." He shakes his finger about an inch from your nose.

"It's a soil incinerator, not a toxic waste dump," you say.

"Same thing!" he exclaims, and you realize that people in the deli are starting to look at the two of you.

"I'm not the enemy here, Mr. Maki," you say. "In fact, I'm looking into just how safe this incinerator will be for the environment. We're doing extensive research."

"Extensive research," Carl says, obviously not impressed. "Ever hear about River Junction?"

"River Junction?" you say. "I'm not familiar with that place."

"It's not just a place," Carl says, as he hands you an envelope. "It's a tragedy, one that's going to happen here." With that Carl stands up and leaves.

You look in the envelope and see a newspaper clipping. You don't see a date, nor the name of the paper. The clipping reads:

> RIVER JUNCTION—Investigators have apparently found the source of the contamination recently discovered in area milk storage tanks.
>
> According to the Department of Natural Resources, the contamination is a result of a variety of factors, all traceable to the Smith Incinerator located five miles upstream of River Junction.
>
> The findings indicate that the Smith Incinerator facility is leaking toxic chemicals into the river and groundwater. These chemicals are ingested by the dairy herds, resulting in contamination. The DNR is also testing the water used to clean out the milk storage tanks, as it may also be contaminated.
>
> Residents are advised to drink and cook with only bottled water until further notice.

You have no idea what to make of this information. When you return to your office, you put in a call to Robert Townsman, a friend of yours from college who works as a researcher at the state Department of Natural Resources. "Robert," you say after exchanging greetings and catching up on small talk. "Can you tell me anything about the Smith Incinerator at a place called River Junction?"

"Not offhand," Robert says, but you let him think about it for a few seconds. "Hey, wasn't that about dairy product contamination? It was in Illinois, I think, about seven or eight years ago."

"What was the deal?"

"The soil waiting to be incinerated wasn't stored properly. Or maybe it was the soil coming out of the incinerator that was doing it, but somehow some pretty toxic chemicals leached into the soil, got into a river and the groundwater, and started showing up when the farmers would test their milk storage tanks."

"What eventually happened?" you ask.

"I don't know," Robert replied. "Send me a formal request for information, and I can look it up for you."

"I'll do that," you say and hang up.

Just then, Jack walks into your office, looking grim.

"What can I do for you?" you ask him.

"Just your job, for one," Jack says, obviously angry. "I hear you've been talking to those goofball farmers."

"Just one," you say, "and he cornered me at lunch today. I'm surprised at how fast you heard."

"Those people are nothing but trouble," Jack says. "Don't believe anything they tell you or that you read in the paper about what they're saying."

You are a little offended that Jack is now dictating what you can and cannot believe. "You know, they may have legitimate concerns. I just learned about an incident in Illinois involving an incineration facility that resulted in contaminated dairy products. I plan on requesting more information from the DNR."

Jack frowns. "We already have enough environmental information to make our impact report." He turns and leaves your office.

You, incredulous, watch him go. You know that Jack does not have the power to keep you from conducting your research—or at least, he's not supposed to. You seriously think about discussing the matter with Bill Donnelly, one of the company's officers. Mr. Donnelly is the only person to whom Jack reports, but you wonder if going over Jack's head would make matters worse. You also really want to know more about River Junction, but you realize that you should get authorization from *someone* to make the formal request to the DNR. You spend the rest of the afternoon drafting various memos to different people, but you still are unsure as to what action you should take.

Suggested Next Steps

- Carefully consider all your options. What different communications could you write?
- Evaluate all the possible effects each of your options could cause.
- Decide on a course of action and write your document.

CHAPTER 4

Technical Definition

Definitions are increasingly important in today's rapidly advancing technological workplace. Imagine, for example, that you were recently hired as a technical communicator at a large computer software company. Your specific activities include writing manuals and advertising copy for a new desktop publishing program that is faster and more user-friendly than any other program on the market. When you write the manual, you must define for users what individual elements mean. If you aren't already familiar with the terminology, how will you define *megabytes, sound card compatibility, CD-ROM speed,* or *platform*? More importantly, how will you translate those specific definitions into general definitions appropriate for advertising copy?

Definitions are also important when companies seek to specify policies or establish legal perimeters. For example, if your company decides to offer employees benefits on the *401K plan,* such a plan must be defined in terms understandable by everyone who qualifies for the plan. If your company develops a new chemical process to protect apple orchards from pests, clear definitions of both the chemicals and the application process will be important for legal purposes.

To define, then, is to state the meaning of a word or series of words. Defining sounds simple on the surface, but *definition* can be a very complex activity, particularly for technical communicators who must write expanded definitions in business and industry. The purpose of this chapter is to familiarize you with three types of definition style: parenthetical definition, formal or sentence definition, and expanded definition. All three are important depending on your purpose and your audience.

PARENTHETICAL DEFINITIONS

Parenthetical definitions are among the easiest you'll write because they consist of little more than a quick (one to five words) definition, in parentheses, immediately following a term that your reader may not understand. For example, suppose you wrote the following sentence:

> The Personnel Manager decided to try a pragmatic approach to the problem.

If your reader understands every word in that sentence except "pragmatic," the reader doesn't understand your sentence. Thus, if you suspect a word needs immediate defining, use a parenthetical definition so that your sentence reads more like this:

> The Personnel Manager decided to try a pragmatic (practical) approach to the problem.

The most important issue, though, involves when and how to use parenthetical definitions. Remember to consider your purpose and your audience.

Purpose

If your communication purpose involves quick definition, use a parenthetical definition instead of a formal or expanded definition. Situations appropriate to parenthetical definitions might include the following:

- Single terms that require immediate definition for reader comprehension
- Technical terms that require clarification for readers undertaking a specific process
- Terms, concepts, or professional jargon that requires translation for a general readership.

Determine your definition purpose before you decide on the type of definition you'll employ.

Audience

Closely connected to purpose is audience, that group of individuals who will read your material. Careful consideration of the expertise of your audience will help you to decide better whether or not a parenthetical definition is appropriate. In general, your audience is likely to fall into one of three categories:

- *General readership.* General readers are those with the least expertise, and thus are most likely to require parenthetical definitions when you are introducing new words or terms.
- *Specialized readership.* Specialized readers are those with some expertise. Such readers might include your colleagues, co-workers, or peers. These readers usually need parenthetical definitions when you are introducing a new word or term or when you want to remind the reader of a definition that is somewhat obscure.
- *Expert readership.* Expert readers are those with the most expertise; thus, such readers rarely require parenthetical definitions unless you are introducing a word or term completely new or highly obscure.

Consider your audience carefully before choosing a parenthetical definition. The needs of general readers are quite different than those of expert readers.

FORMAL DEFINITIONS

Formal definitions are sometimes called *sentence definitions* because they are complete sentences that offer more information about a term than does a parenthetical definition. If a parenthetical definition provides readers with a quick reference point, the formal definition gives readers a more complete understanding of the term by providing, the *class* into which the term fits and the *unique features* that distinguish that term from all others. Notice in the following example the three component parts of the formal definition—term, class, and unique features:

Term	*Class*	*Unique Features*
Spark Plug	a device	that is screwed into the end of each cylinder of an internal combustion engine, containing two electrodes between which an electric spark jumps that ignites the fuel.

Notice that the formal definition most closely resembles the type of definition you might find in the dictionary—formal, complete, and thorough. Consider the following additional examples:

Term	*Class*	*Unique Features*
Syringe	an instrument	for injecting liquids into the body or withdrawing them from the body through a narrow tube, operated by a rubber bulb.

Escrow	a sealed instrument	given to a third party for delivery upon stated conditions.
Hypnosis	a state	resembling sleep, often artificially induced, in which the mind readily responds to external suggestion.
Rudder	a vertical fin	at the rear of a vessel or aircraft used for steering.

All of these definitions provide the reader with a good sense not only of the meaning of the term, but also of how that term fits into a larger group or class of items. As was the case, though, in parenthetical definitions, use a formal definition after you have considered both your purpose and your audience.

Purpose

Use the formal definition if your purpose is to provide a more complete definition of a term. For example, if you are writing a manual to accompany a new computer, you might include formal definitions for terms like *function keys, macro, merge,* or *escape.* In such a case, readers need more information than a parenthetical definition can provide. Formal definitions, in addition, provide very specific information. This is particularly important if you want to make certain readers don't mistake definitions. The term *escape,* in a computer-related context, means something very different than does the term *escape* in a kidnapping case.

Audience

Consideration of the needs of your audience remains very important, so determine the expertise of your readers before settling on a formal or sentence definition. The following types of readers have very different needs:

General Readership

General readers require the most information, so sentence definitions are particularly useful when you are trying to provide complete specificity. General readers, for example, may find it very confusing if you demonstrate the unique features of a term by using the term itself.

> A libertarian is a person who believes in liberty for all people.

This formal definition is *circular* because it defines the term (*libertarian*) by using essentially the same term (*liberty*) to show unique features. Use, therefore, terms that clarify the definition.

> A libertarian is a person who believes in freedom of conduct or thought.

TIP

> Since you will likely create expanded definitions for a more general readership, consider sharing your rough draft with a family member or friend. Often readers unfamiliar with a topic can provide useful feedback during the drafting stage.

Specialized Readership

Specialized readers also have particular needs, though their level of expertise is greater than the general readership. Thus, specialized readers find formal definitions especially useful when the definitions show precise distinctions between words or terms. One way to increase specificity in your formal definitions is to narrow the class into which your term fits. For example, if you define a *wrench* as "an object used for grasping and turning nuts," your reader may wonder what kind of "object" you are defining. A better formal definition might read like this:

> A wrench is a tool used for grasping and turning metal machine nuts.

While specialized readers may have some expertise, they also benefit from specificity.

Expert Readership

Expert readers, as we discussed earlier in this chapter, bring to your document the most background knowledge. Formal definitions, however, are still useful for such readers. Just as specificity in the class is a useful feature for specialized readers, so specificity in the *unique features* portion of your formal definition is useful for expert readers. Expert readers generally already know a good deal about your topic, so providing them with precise unique features helps them to make the fine distinctions in definition that are less important for, say, general readers. For example, if you define a *tort* as "a civil wrong," you don't provide an expert reader the same specificity that the following definition provides:

> A tort is a civil infraction, typically a negligent or purposeful act resulting in injury to a plaintiff's property, person or reputation.

You may argue that an expert reader likely already knows the definition of the term, but a narrow, precise presentation of unique features helps even the expert reader derive the greatest level of understanding from your formal definition.

As you can see, formal definitions are a useful means of explaining a word or term for a variety of purposes and audiences. In fact, formal definitions often constitute the opening definition in an *expanded definition,* the longest and most complex of the definition types we will analyze in this chapter.

EXPANDED DEFINITIONS

The expanded definition is the longer, more detailed, paper-length version (usually 500 to 1,000 words) that you will write either in response to one of the following cases or in response to your instructor's specific assignment. The purpose of the expanded definition is to present to readers a thorough, complete understanding of an item, concept, or mechanism. Expanded definitions typically contain a variety of different parts into which you break down and evaluate whatever it is you are defining. Those parts include, but are not limited to, the following:

Introduction

This usually includes history, background material, and the sentence definition.

History and Background. Readers find it useful to learn a little about the history or background of your item or concept, so this is a good opportunity to do a little research either in the library or on the Internet. For example, if you are defining the low-pressure, water-saving, contemporary version of the standard toilet, your readers might find it interesting to learn that the inventor of the device was named Thomas Crapper!

Use the introduction as an opportunity to try the sentence definition as well. Readers find it useful to see the item or concept in the context of term, class, and unique features.

Body

The body usually includes analysis of parts, exemplification, required components or special conditions, negation, and typically at least one visual.

Analysis of Parts. Before the readers can understand the item or concept you are attempting to define, they need some understanding of the parts that make up the whole. For example, as you will see in the sample paper following this section, to define *dactyloscopy* (the science of positive identification of a human through fingerprints), you must discuss the parts that make up the human fingerprint.

Exemplification. Exemplification means to "show by example," a very important aspect of definition. You assist the reader in understanding your item or concept by presenting examples that support your definition. A definition paper on the gasoline engine, for example, might include early examples (a Model T engine, for instance) to more contemporary, fuel-injected models.

Required Components or Special Conditions. In the process of defining your item or concept, you also need to explain to readers what components or special conditions might be required. For example, if you are defining *autopsy,* you will probably also need to explain what implements (scalpel, Stryker saw, and other surgical tools) would be used by someone performing an autopsy.

Negation. Negation, very simply, is the act of telling your reader what something is not. Readers benefit just as much from such information as they do from comparisons. For example, in the *dactyloscopy* paper following, the "fork" and "spur" appear similar but are different. Explaining how a "fork" is not a "spur" will help the reader make a potentially important distinction.

Visuals. Visuals or graphic aids are especially useful when you are defining. Because definition is a difficult exercise, the incorporation of visuals into your paper increases potential reader comprehension. Notice the visual used in the following *dactyloscopy* paper. Without the graphic, readers may have difficulty visualizing the various parts of the human fingerprint. Thus, use visuals in your own expanded definition paper. Refer to the appendix of this textbook, Creating and Working with Visuals, for more information on using visuals in technical writing.

Conclusion

The conclusion provides information on the operating cycle and gives your summation.

Operating Cycle. Virtually anything you might choose to define has some kind of operating cycle. You really could not, for example, define *floppy disk* without explaining how it works in a computer. Thus, most expanded definition papers conclude with a discussion, when relevant, of the operating cycle of the item or concept being defined.

Summation. In addition, most definition papers conclude with some kind of summary of the entire definition. Think of the summation as an opportunity to provide readers a final note or an additional sentence or two of information that completes your discussion of the item or concept.

TIP

Avoid putting new information in your summation. It is good to repeat important information, but do not save your best point for last.

All of the parts of the expanded definition are important, but keep in mind two important points.

- Depending on your topic, not all the parts may be relevant to your paper. If you are defining something truly unique, negation might not be useful to the reader.
- You may decide on an alternate order for the parts just discussed. You may, for example, decide to conclude with historical background rather than open with it. The suggested arrangement of parts is just that—a suggested arrangement.

For further clarification of the expanded definition paper, consider the following student-written definition of *dactyloscopy.*

DACTYLOSCOPY

Formal definition

Dactyloscopy is the science of positive identification of a human being through fingerprints. Because no two identical prints, even of individual fingers, have ever been found, a matching print pattern can conclusively identify an otherwise unidentifiable deceased person, convict a suspect whose prints were found at a crime scene, or unmask individuals assuming a fictitious identity.

Historical background

Since the establishment of the Galton and Henry print identification systems (circa 1900), using characteristics of individual ridge lines within a given area, the uniqueness of every finger on every hand on every human being has increased so exponentially that the discovery of two identical prints is now viewed as a mathematical impossibility.

continued

Analysis of parts

Human fingerprints fall into three basic patterns: *loops,* which constitute about 65% of all prints; *whorls,* about 30%; and *arches,* about 4%. *Anomalies* make up about 1%. Loops and whorls have a central focus called a *core.* Where individual ridge lines converge near a point, the arrangement is called a *delta.* Whorls have a core and two deltas. Arches have neither cores nor deltas.

Exemplification

A single ridge line can contain small proturbances called *hooks* or *forks* if leaning toward the right, or *contra hooks* or *contra forks* if leaning toward the left. Ridge lines that split into two new lines are called *bifurcations,* and between ridge lines may be found other small phenomena termed *eyes* or *islands,* some solid and others blank. Figure 4.1, shows the parts of a fingerprint.

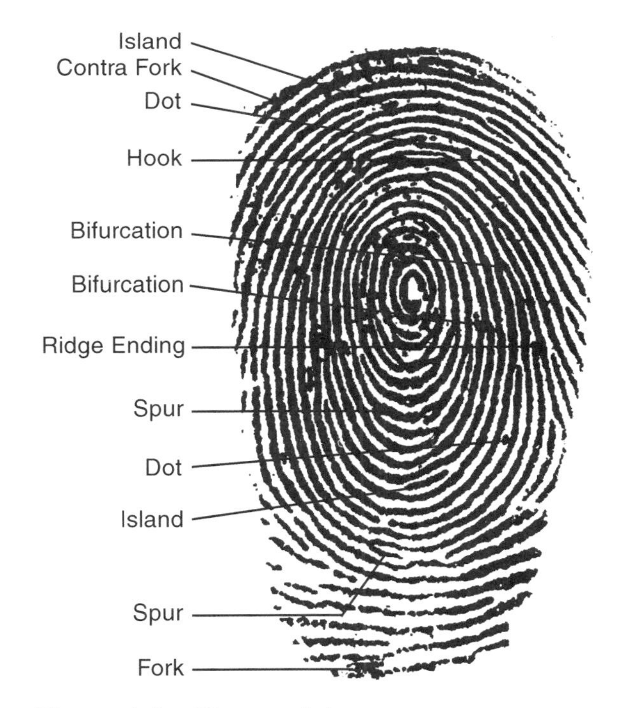

Figure 4.1 Fingerprint

Special conditions and operating cycle

Fingerprints may be left at a crime scene in three ways: **visible,** as in a smeared or bloody print; **plastic,** a concave

continued

depression in butter or candle wax; or **latent,** a nominally invisible print. By brushing latent prints with fine powder or chemical sprays, the print becomes visible on a wall, glass, furniture, or virtually any flat surface. This is because of the residue of human acidic perspiration continually secreted in the ridge lines. When this oily substance is dusted or sprayed, a definite print pattern is revealed.

Summation

Multiplying three basic patterns times the approximately twenty ridge lines per finger times ten fingers and hundreds of variables among all the ridge lines times six billion human beings, the distinctions in dactyloscopy approach infinity, thus rendering indentical prints impossible, and establishing, therefore, an absolute indentity for every individual.

Exercises for Classroom Discussion

1. You are the meteorologist for a local television station in a small midwest community. One aspect of your job you have come to enjoy is interacting with the public to explain meteorological phenomenon in either community lectures or at the local high school. Because *global warming* has become a popular topic in both the scientific and popular press, your boss, the station manager, thinks of you when the local Lion's Club president contacts him looking for a guest speaker who can address the issue of increasingly hot springs and summers in the Midwest. You discuss it with your boss and decide to present, in one week, a talk on the weather-related phenomenon known as *El Niño.* Because your audience knows little about the topic, you must first define the phenomenon and then explain how it causes changes in the weather pattern. You decide to check the Internet for a complete definition and for some information on the topic.

2. You are the assistant prosecuting attorney in a midsized Utah city (population 40,000). Your case against a suspect accused of assault with a deadly weapon is strong, based on the ballistics evidence. Your colleague, the lead prosecuting attorney, is concerned, though, that the jury of twelve largely uneducated men and women will have trouble convicting the suspect based solely on ballistics evidence. Because the evidence, and more importantly the interpretation of the evidence, is critical to the case, you approach the judge for permission to clarify the term *ballistics.* After a sidebar conference with the judge, you are granted a ten-minute opportunity to define the term *ballistics* at 8:00 a.m. the following morning. Your case may hinge on the jury's understanding of this term, so you decide to stay up that night writing a careful definition of the term for a nontechnical audience.

3. You have recently been hired as the Chief Executive Director for the Department of Agriculture in Wisconsin. Because of government downsizing, your district consists of several small rural communities that for many years had been served by onsite agricultural offices. The new, centrally located office that you will head is no more than 40 miles, one way, from any of the communities you will now serve. Because of the distance farmers will have to travel to visit your office, you decide to put together a presentation, to be held in a school gymnasium, on several new programs funded by the USDA and invite the 170 farmers you serve to attend. Because a large percentage of the farmers consistently grow corn and some soy beans, you decide to promote the USDA program on *sustainable yield* as a way of introducing new crops to the area. Selling these farmers, many of whom have been in the business all their lives, on the sustainable yield program will not be easy, so you decide to write up a solid definition of the practice, with plenty of examples, for presentation in one week.

4. You are a new nurse assigned to St. Mary's, a local hospital serving a community of about 25,000. In addition to your regular responsibilities with patients, you are expected to visit the local nursing home, Mulberry Heights, once a month to assist the medical support staff and address issues of health with the elderly residents. You are scheduled to visit the nursing home in just a few days, and your supervisor asks you to make a brief presentation on *cholesterol,* what it is, how it is controlled, and what effect it can have on the elderly. Because your audience has a low level of expertise, you decide to define the term, provide several examples of foods high in cholesterol, and suggests ways in which residents can monitor their own cholesterol intake.

5. You live in a suburban neighborhood on the outskirts of a growing city. The city's tax base has increased for several years because of a variety of new businesses—from an automotive plant to a subsidiary of a software giant from the West Coast. As a result of the growth and the increase in population, the city manager has proposed creating a new airport, which would service both commuter and jet aircraft, approximately 15 miles from your suburban neighborhood. While city officials are excited about the potential of the new airport, you and a small group of concerned citizens are less than excited about the potential noise pollution that jet aircraft would bring to the area. You, along with a few neighbors, decide to bring the issue to the city commission, but you need facts to support your case. You decide to do a little research on *decibel ratings,* define the principal and establish the ratings of jet aircraft. You believe that the decibel ratings will be high enough to cause permanent hearing loss, but you've got to convince a number of people who really want the new airport.

SCENARIOS FOR TECHNICAL DEFINITION

The New Technology Problem

CONSIDERATIONS

In this situation, it's your job to help your company identify and learn about cutting-edge technologies. This information will be useful to Matheson & Matheson's investment team as they make their predictions as to what companies will provide the greatest return on investments. As you read this scenario, keep the following questions in mind:

- What kind of information are you being asked to find?
- Who are all the potential readers of your document?
- What new products and technologies have you heard or read about?

It's your first week at Matheson & Matheson, an investment firm in Sacramento, California. You just completed your corporate orientation, and you are excited to begin work on your first research project, which will be discussed at this morning's meeting.

"Hello," Jenny Mathias, research director, greets you as you walk into the meeting room. "I hope you're ready for some long hours on the Internet."

"Sure," you say taking your seat. "What would you like me to research?"

Your colleagues around the table smile. "Am I missing something?" you ask.

"Not really," says Daniel Birch, your manager. "We really don't know what we want you to research."

"You see," Jenny continues, noticing your blank look, "we are in need of some new, fresh ideas. We have been buying stock in the same companies for the last few years, and we want to pursue some new, more lucrative investments."

"Understandable," you say. "What new companies are we looking at?"

"Well," Daniel interjects, "that's where you come in."

Later, in your office, you sit down at your computer and reflect on the meeting. The discussion mainly concerned the firm's desperate need to find a small-scale client with enormous growth potential. In fact, it seemed to you that if Matheson & Matheson didn't find some new high-yield investments, you might be looking for a new job again.

It's obvious to you that the kinds of companies with the largest growth potential would be connected with high technology. The amazing growth of companies such as Microsoft and Netscape made fortunes for many investors. Your team's job is to find a new product or innovation that may have serious growth potential, and then provide the investment board with a collection of small reports, from technical definitions to market analysis to company history, current investor information, and so on.

Your part of the project may be the most delicate. As the newest employee, your team wants to see the caliber of your work and, therefore, has asked you to do the research and come up with a technical definition of a product or service you believe to be unique, relatively unknown, and in need of investors.

You turn on your computer and begin logging on to the Internet. *What could I possibly find?* you think. *What's new in certain fields—computers, medicine, telecommunications, entertainment, recreation, transportation, finance . . .*

Your mind jumps around as you find a good search engine. You pause for a moment, wondering what to enter into the "Search" field. You have just a few days to write your definition, and you want to settle on something to define as soon as possible.

Suggested Next Steps

- Review the guidelines for creating technical definitions as outlined in this chapter.
- Decide on a product or service to research. Collect all the information necessary for a technical definition. Be sure to document the sources for your information.
- Refer to Creating and Working with Visuals (the appendix of this textbook) as you prepare at least one visual for your technical definition.
- Think through your reasons for choosing this product or service to define. Be sure to discuss and explain your choice in your Solution Defense.

Cable Modem Pilot Project

Considerations

As a technical writer for a cable television company, it's your job to help people understand cable and television technologies. In this situation, you are investigating a new type of equipment that will bring Internet access into the homes of people who have cable TV. When reading this scenario, keep the following questions in mind:

- What kind of information are you being asked to find?
- What purpose will your document serve?
- Who are your potential readers, and what is their level of understanding of this new technology?

You are a technical writer for Salem MassCom, a small cable television company serving the New England area. Today, John Little, Vice President of SMC, has asked you to meet him in his office. Mr. Little is sitting behind his desk with his back to the door when he calls you in. He turns his chair around and smiles from behind wire-framed glasses. You notice an unusual black abstract sculpture on his desk and comment on it.

"It looks like wings on fire," you say touching the cold stone.

"Ah, yes," he says. "It's by a promising young artist friend of mind. 'Inspiration' he calls it. He presented it to me to remind me that I need to keep from being complacent."

"You feel you're in a rut?" you ask, a little surprised by Mr. Little's openness.

"Maybe," he says, smiling ruefully. He rubs his hand over his bald head. "Technology moves so quickly, and I was struggling to keep up with it when I became Vice President of SMC. Now, I'm so busy I don't find enough time to see what's new out there."

You nod, waiting for him to get to the point. "Have you heard of cable modems?" you ask.

"A little," you answer.

"What do you know about them?"

"Just that they are one hundred to one thousand times faster than a standard phone-based modem," you say. "And that you can stream video and audio over the Internet on them, which is really slow on phone modems. Other than that, I haven't been doing any research on the technology."

"Well, we're considering doing a pilot project in Maine with cable modem Internet access," Mr. Little says. "It's possible that television and computers will merge into the same thing, and we need to be ahead of the game. But I know practically nothing about the technology and neither does Carla Patrick."

You are a little surprised that Ms. Patrick, President of SMC, is not up on new technologies. She recently won an award for her Cable and Education programs. You have personally provided her with several technical documents on satellite technology.

"What kind of information are you looking for?" you ask.

"Well, the basics of the technology, how long its been around, what we can do with it in terms of Internet applications, and any problems other companies using it have had," Mr. Little says. "And I need it as soon as possible."

"I'll get right on it," you say. "Anything else?"

"No," he says. You stand to leave. "Wait just a second." You turn to find him standing with the odd statue in his hands. "Why don't you put this on your desk while you work on this," he says. "I want you to know how important your research is going to be to us. If we invest in a cable modem project, SMC just might become the leading telecommunications company, instead of just a cable company. This could make or break us."

You nod, take the statue, and leave.

Later, you sit in your office, looking at the "Inspiration" statue. Your computer is starting up, so you have a few seconds to think. Finding information on cable modems will not be hard—the Internet is probably loaded with sites about the technology. You wonder, however, how to work the definition so John, who is obviously a low-tech kind of guy, will feel comfortable with the information. You also know that your documentation will more than likely be read by Ms. Patrick, and your experience with her was such that you know she doesn't like something over-explained to her.

Your computer makes tonal music, indicating it's started up and ready to roll. You type in your Internet log-on ID and begin searching.

Suggested Next Steps

- Conduct research, using the Internet, technology-related magazines and journals, and other sources to gather information on cable modems.
- Review the guidelines for creating technical definitions provided in this chapter.
- Carefully consider your main reader and any other potential readers. Be sure to include your rationale for how you worded the technical definition in your Solution Defense.

The Pitted Park

Considerations

In this situation, you are responsible for alleviating fears about a new treatment for crumbling stone statues. As you read the scenario, keep the following questions in mind:

- Who are the potential readers of your document?
- How will readers use your document to make decisions?
- What kind of information do you have for this project, and what kind will you need to complete your document?

You work as a researcher for Preservation, Inc., a small company that tries to restore outdoor statues damaged through vandalism, accidental damage, and the normal decay from environmental factors.

Your current project is in Penbrook, a small town with a dozen limestone sculptures in its Bicentennial Park. The park, created in 1976 to commemorate the nation's 200th birthday, is located in the center of town.

The sculptures of famous Americans, from George Washington to John F. Kennedy, are pitted and some have areas so brittle they crumble when touched.

Bonnie Cooper, your partner at Preservation, Inc., comes into your laboratory/workshop just as you are finishing smoothing plaster over a gaping hole in a stone bird's wing. You will use the dried plaster to make a mold, which you will then use to create a plastistone section of the wing. Attaching the plastistone to the real granite is, for you, the harder part, but you want to talk with Bonnie before you go any further.

"How'd it go?" you ask, referring to the city commission meeting Bonnie had attended the night before.

"Good news or bad news?" she says, reaching for the coffee pot. "This is coffee, right? You haven't been cooking glues in here again?"

You laugh and assure her the liquid is, indeed, coffee.

"Good news first," you say.

"They approved our budget as proposed," she says.

"Excellent," you say. "Can't wait to get started with the new stuff."

"That's the bad news," Bonnie said, stirring cream in her coffee. "I sat there for half an hour listening to Bob Biltmore rattle on about how the great masters' paintings were ruined when people, in an effort to restore them, painted varnish on the canvases."

"So he thinks AEPS is like varnish?" you ask incredulously. "Don't tell me they won't let us use it. I don't know how else to keep the statues from disintegrating."

"Oh, I think they'll let us use it," Bonnie says. "I need you to write up an explanation of AEPS—what it is and how it works. You know, define it for them. They want to know exactly what they're putting on George Washington's nose."

You laugh. "I'll get right on it," you say.

"I know you're a good writer," Bonnie says, "but I just want to caution you. None of these people knows a whole lot about chemistry. Try not to fly over their heads."

"I'll be good and grounded," you say turning back to the bird statue.

As you continue to work on the statue, you think back on how you first learned about aminoethylaminopropylsilane, or AEPS* for short. You had been working for Penbrook on another project—the deterioration of the city's limestone-faced courthouse.

A colleague had sent you an article about a project between the New York Metropolitan Museum of Art and chemists at Sandia National Laboratories. The restorers at the museum had been trying to use a chemical that works like liquid glass, but it would not adhere to the limestone. Limestone

*The source of information on AEPS and statue restoration is "Breakthroughs." *Discovery Magazine,* June, 1997.

is basically made up of calcium carbonate, and AEPS seeps into the cracks in limestone and binds with calcium carbonate. The liquid glass then sticks to the AEPS, thus protecting the statue. The finished process resulted, according to the article, in treated stone with ten times the resistance to deterioration than untreated stone.

You decide to do some formal research into some of the other techniques in use before the AEPS and liquid glass treatment. Perhaps that, along with an explanation of the process, will convince the council that AEPS is safe and effective.

Suggested Next Steps

- Research stone statue restoration techniques in order to put the AEPS treatment into context.
- Review the guidelines for creating technical definitions as provided in this chapter.
- Create a visual that illustrates the AEPS treatment process. Refer to "Creating and Working with Visuals" at the end of this textbook for guidelines on visuals.

The Integrated Ed Problem

Considerations

As a teacher, you are constantly enhancing your own skills to find newer and better ways of imparting knowledge to your students. In this situation, you are asked to educate your colleagues about a new teaching method called "integrated education." While reading this scenario, keep the following questions in mind:

- What are the aspects of integrated education, as you know it?
- What are the concerns and motivations of your colleagues concerning this project?
- What additional information do you need to fully understand and explain this teaching method?

You are an English teacher at Second Chance High School, an alternative education school for students who could not function well in the traditional high school. Most of Second Chance's students have behavioral problems, and some are learning or emotionally impaired. In addition to teaching English, you also sit on the school's curriculum committee. The committee discusses what methods to use when teaching and decides on the content areas and issues to be covered in every subject for each grade.

During your planning hour, Rebecca Simms, the principal of Second Chance, comes into your room. You ask her what you can do for you. "Well, I'd like you to prepare a document for the Curriculum Committee," Rebecca says. "We need to begin our discussion of integrated education and how it would work for us."

"I agree," you say, flattered that the principal remembered your interest in integrated education principles.

"Why don't I prepare a definition of the concept, so we all understand exactly what it means and involves?"

"Exactly what I was looking for," Rebecca says. "I am particularly concerned about the committee's reaction to having to prepare new course materials each semester. And the division of responsibility from teacher to teacher needs to be addressed. That's the biggest objection that I've heard so far. There are a lot of little ones, but they aren't as significant."

"What were those 'little ones,' if you don't mind?" you ask.

"Well, there was some argument over what subject would lead into the others. So far, there are two camps—that history should lead all the subjects or that literature, taught chronologically, should lead."

"There's really no such thing as 'leading' in integrated education," you say.

"Well, you need to explain that," Rebecca answers. "I'd like to have this definition by the next meeting."

"No problem," you say. She says good-bye and leaves your room.

You sit and ponder how to approach this assignment. Changing from a method of teaching by strictly divided subjects to integrating all subjects would be a major challenge. In your opinion, though, a school like Second Chance needs such a change. Students have a hard time seeing how history, literature, science, and mathematics all relate in many ways.

You think back to your college years and the integrated education retreat you attended. The retreat was set up like a school field trip. The teachers-in-training posed as students, who participated in a variety of activities while on a "nature hike." In addition to the expected lessons on plants and animals, a master teacher talked about the historical significance of the Native Americans who inhabited the area. The group took water and soil samples, wrote about their thoughts and feelings in their journals, and received astronomy lessons while sitting around the campfire. They also learned some of the traditional "fireside" songs sung by Scottish farmers near the turn of the century. Once back in the classroom, they ran the data from their water and soil samples through a computer and published their results on an Internet web site. One group studied the water samples under the microscope, taking a count of a particular kind of protozoa found. Some of the stories from the journals were also published after a traditional workshop session and some revisions.

For you, what "integrated education" really meant had not been very clear before attending the retreat. Essentially, the teaching concept involved taking down the divisions between subjects and allowing students to form more connections between what they study. The idea is to also show students that being "good" at just one or two subjects would not help them succeed in a world where all kinds of different knowledge types are required.

You reflect on your conversation with Rebecca. Integrated education often seemed like teaching within controlled chaos—sometimes lessons are built around time periods, sometimes around themes. For Second Chance High, switching to an integrated curriculum might do more than help teachers reach more students—it may provide the school with some measure of prestige as an experimental teaching institution. The school operates strictly on grants and a limited amount of state and federal funding, so any opportunity for special recognition could only increase their chances for larger operating budgets.

Because it has been a while since you were completely up to date on the idea of integrated education, you decide to do some research on current trends and case studies before writing the definition. You think about the arguments from faculty who do not want to change the way they've taught—some have been using the same tests and lectures for over a decade.

Suggested Next Steps

- Conduct research on the latest developments and assessments of integrated education. Use the Internet and professional journals, and be sure to investigate papers presented at conferences such as the National Council of Teachers of English (NCTE).
- Review the guidelines for creating technical definitions, provided in this chapter.
- Carefully consider the fears and self-interests of your potential readers. Discuss how you addressed these issues in you Solution Defense.

Soil Incineration: Defining in Muddy Waters

Considerations

In this portion of your work on the soil incineration project, you must create the first of what you know will be a series of technical documents. While reading through this situation, keep the following questions in mind:

- Who are your readers for this document, and what purposes will the document serve for them?

- Is there information from your prior work on the project that impacts the creation of this document?
- How will you deal with the incineration process problems?

It's coming down to crunch time for you at Donnelly Engineering. Jack MacGregor has given you a pretty short deadline to come up with a definition report for the soil incineration project. You've been juggling three other projects, and your intern has left for a well-deserved vacation. The county board needs the definition for its upcoming meeting, and you have just a few days to prepare the document. Jack has already warned you that POL members will be presenting their own documents in hopes that the county will stop the project in the preliminary phases. From your understanding of the county board, they have little prior knowledge about soil incinerators. The only related issues, according to research your intern performed, had to do with a three-month-long fight with the state to allow the county to transport fuel-oil contaminated soil over state lines. You vaguely remember a pile of dirt under a blue tarp that sat outside the area's largest gas station for most of the winter and the editorials in the newspaper about the county's inability to dispose of the soil.

You clear your schedule for the next hour and start gathering the information on soil incineration processes in general. The final decision for how this particular plant will be constructed had not yet been made—the client was still considering a few options concerning dust collection.

You look at the basic contaminated soil plant flow diagram (Figure 4.2) and start thinking about how to define this process. From what you understand of soil incineration, the soil isn't really burned—more like baked to an extreme degree. The incinerator contains a large drum that holds up to fifteen tons of soil. The drum is turned by a large motor, and heated by natural gas to seven hundred degrees Fahrenheit. The remaining soil is then free of toxic chemicals, but the gasses released when the soil is baked are the problem. The gasses are run through several air collection systems, which grab any airborne particles. You read the description of these systems and wonder how you will explain them to the board: cyclone-particulate removal, baghouse-particulate removal, and afterburner-voc combustion.

You know that a cyclone-particulate removal system involves using a screw conveyor to grab the particles and pull them into the center of a hopper, where they are gathered and then moved to a storage bin. The baghouse is a system of special filter bags on all the drum ventilators that do not allow any particles to escape. The incinerator operators must change the filter bags periodically—just as you would a vacuum cleaner bag. The afterburner is slightly more complex. According to the material your intern provided, the air from the drum is passed over a burner flame, which oxidizes the contaminants (volatile organic compounds [VOCs]) into carbon

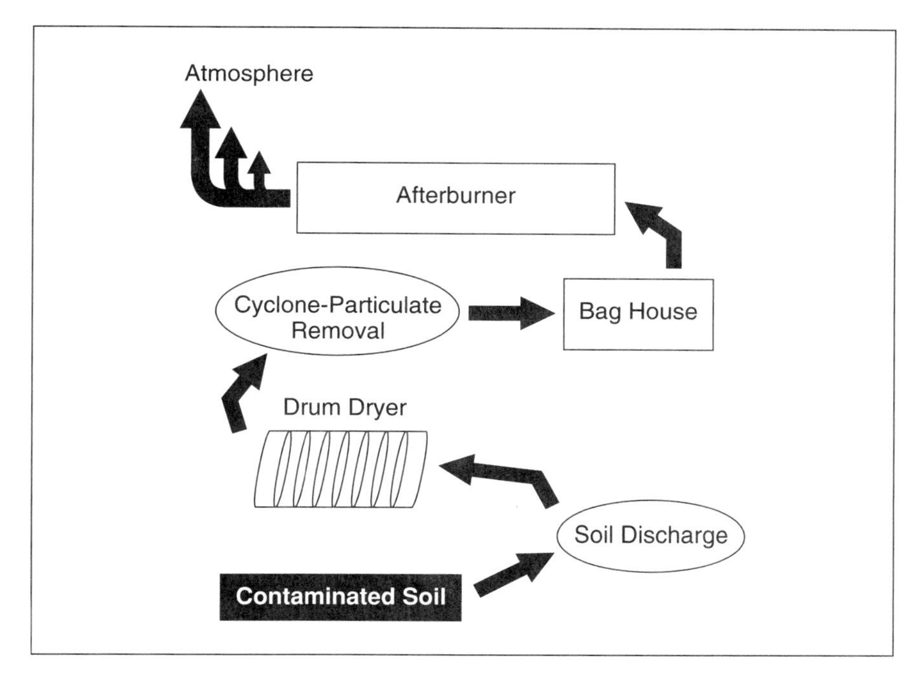

FIGURE 4.2 Contaminated Soil Plant Flow Diagram

dioxide and water. Temperatures in the burner area can be up to 1,800 degrees Fahrenheit, which means a great deal of fuel consumption.

Beyond the basics of the plant's operation, you know that the POL organization will have plenty of information about the drawbacks of soil incinerators. You and Jack have already discussed the problem with keeping the contaminated soil from blowing around when it's transported to the drum. One solution would be to wet the soil down, but there are two consequences. The water running off the soil will be contaminated, and the drum will not efficiently burn damp soil. Storage of the contaminants left over after the collection process is less of a problem; federal and state laws require them to be placed in airtight plastic drums immediately after collection. Another issue is the "clean" soil coming out of the plant. Decontaminated soil still contains heavy metals and other potentially harmful materials. Jack, of course, asked you to "downplay" these issues for the county report.

"I don't want to give POL any ammunition," Jack said sternly.

"We need to be honest about all the effects of this process," you replied.

"Whose side are you on?" Jack had demanded. You quickly backed down and assured him you wanted to do nothing more than provide all the information necessary to keep people from thinking Donnelly Engineering had anything to hide.

You sigh and begin writing a brief technical definition that will accompany Jack's longer proposal arguing for approval of the site.

Suggested Next Steps

- Review the technical information provided in this scenario, and decide how you will define the incineration process. You may want to conduct outside research.
- Consult the guidelines for preparing technical definitions provided in this chapter.
- Consider any ethical dilemmas indicated in this scenario. Discuss how you address these issues in your Solution Defense.
- Carefully analyze your potential readers. Discuss how you will address their concerns and knowledge level in your Solution Defense.

CHAPTER 5

Technical Description

For a variety of reasons, the ability to describe an object or concept is one of the most important skills you can have in business or industry today. Think of the number of times you have said or heard someone else comment, "I have a great idea, but I just don't know how to describe it." This chapter will familiarize you with some of the ways in which *technical description* is handled in the workplace. In industry, technical writers descriptions of physical objects, mechanisms, and concept constituents. Imagine, for a moment, that you have just been hired as a technical writer for one of the big-three auto companies in this country. You've had plenty of classroom experience but little practical experience. Your first job is to write a portion of an automotive manual *describing* the new antilock braking system that will be a standard feature in all new vehicles. How will you go about researching the mechanism involved so that you can write a description that will be used, presumably, by consumers?

Descriptions are also important for legal purposes. A blueprint of a skyscraper is a kind of description that, if drafted with errors, can result in structural problems. The federal Food and Drug Administration carefully monitors the *descriptions* of pharmaceuticals so that consumers are well informed about the medications they purchase. Some companies, such as the one depicted in Figure 5.1, put descriptions and specifications in advertisements for their products. Descriptions are, thus, a very important aspect of both the workplace and our everyday life.

PREPARING TO WRITE A TECHNICAL DESCRIPTION

To describe is to portray or depict in speech or writing. Describing, you may argue, seems like *defining,* and if you notice similarities between the two, you are correct. Describing and defining do share certain characteristics. The important distinction between the two, however, has to do with your ability to help the reader *visualize* the object, mechanism or concept. A good description will help a co-worker, a consumer, or even a competitor "see" what you have in mind. The emphasis on "seeing" your item or concept also makes description a good place for visuals and figures that will aid the reader. You may also be thinking that description seems too simple for an essay-length writing. Sometimes a simple description may be brief.

> *Shrouds,* resembling telephone wires, are cables used to support the top portion of the mast with the sides of the hull.

In most cases, however, a good, thorough description must be longer to allow for adequate detail. On the surface, for example, a mouse trap may seem a simple tool requiring a simple description. Consider, though, all the parts that make up the mousetrap: the wooden plate, the spring mechanism, the bait plate, the hook, and so on. By the time you isolate all the parts and name them, you will likely write a long description of the mechanism. Preparing to write an extended description, then, means limiting the scope of your project, concentrating on the observable detail you will record, naming those details with as much specificity as possible, and considering the needs of your audience.

Limiting the Scope

When you first decide to write the technical description, you may find yourself selecting something to describe that is simply too broad in scope. For example, if you decide to describe a *computer,* you then imply that you will describe the whole computer. Think about all the parts that make up a computer, and you'll realize quickly that such a project would be far beyond the requirements of an extended description. In fact, such a project would likely be more in keeping with a set of specifications (a statement of the dimensions or characteristics of an appliance or apparatus) for the workplace.

For the purposes of the extended description, narrow the topic until you have selected a manageable portion of a larger whole (e.g., the carburetor of a gasoline engine) or a smaller project for which the whole is workable (e.g., a spark plug or a radiator). Thus, choose a topic for your technical description that you know well, contains several parts for analysis, and is limited in scope. Once you have chosen a good topic, you must begin recording observable details about that item.

Recording Detail

Technical description is, in some ways, unique from other forms of technical writing in its emphasis on visual detail. Readers of your description—sometimes a general readership like consumers or sometimes a specific readership like colleagues—will rely on those details that facilitate understanding. As a result, select observable and concrete detail. The best way to record detail is to think of your item or mechanism in terms of an exploded visual (see *Guide to Creating and Working with Visuals* at the end of this text). Imagine that your item or mechanism has been blown apart into separate pieces, and then put those pieces back together again. Think of each piece as a separate item and record information one piece at a time.

Keep your records factual and accurate; do not "color" your observations with personal feelings or adjectives that do not literally describe but instead add *qualities* to your item or concept. For example, imagine that you are a state health inspector who evaluates a chicken processing plant. You must write a report on your visit, describing the conditions at the plant. You find the conditions to be far below state-certified guidelines. Decide which of the following two descriptive report portions would be acceptable.

Report A

Conditions present in the cutting and sorting area were the source of concern. Inadequate drainage caused poultry blood to pool on the cutting tables and occasionally drip on the floor. Insects were present, as was fecal matter on the sorting tables.

Report B

Conditions present in the cutting and sorting area were really awful. Blood was everywhere and the sickening smell took your breath away. Flies were buzzing around the sticky mess. The whole thing was just appalling.

Remember to record what you observe but in a manner that maintains the details you want your audience to see. A supervisor reading through the segment from report B only knows that the individual in charge was upset by the conditions. The supervisor is much more concerned with a visual picture of the conditions observed and evaluated.

Naming Detail

This section may seem obvious—naming the detail you observe—but in actuality the act of attributing names or "tags" to observable detail may be difficult because not every part of every item you observe has a recognizable "tag." Let us go back to the mousetrap example earlier in this chapter. Many of us have observed and maybe even used a mousetrap, but how many of us have thought about what the parts of the mousetrap are called? Certainly the wooden base of the trap is just that—a base—but what about the thing that holds the bait? What about the little piece of metal with a slight protrusion that keeps the wire lever in place? These simple examples demonstrate how the technical writer must sometimes name or tag these parts for the audience and for the purposes of any visuals that may be created.

Regardless of what you decide to describe, remember to select useful, concrete names or tags for parts. Do not refer to "the tiny doodad that holds the lever in place" or "this 'thing' that stops the heating element when the light comes on." If you don't have a name for a part, invent a good name that expresses the purpose of the part. That thing that holds the bait on the mousetrap? Why not call it the "bait plate"—a name that expresses what it is and what it looks like.

TIP

Avoid stacking too many noun forms when you are naming items. For example, if you are describing a TV satellite dish, you might confuse the reader if you call it a remote, multifunction, directional television transmission receiver.

Considering the Audience

Like all technical writing assignments we've covered so far, the technical description demands attention to the needs of the audience. Like the definition paper, for example, the needs of the audience depend greatly on the expertise of the audience. Generally speaking, you will be writing for audiences with a high level of specialization (experts), a moderate level of specialization (colleagues or competitors), or a low level of specialization (consumers or the general public). Let us consider, as an example, what experts, colleagues, and consumers might need in a description.

Expert readers can handle the highest level of specificity and detail. If you are a technical writer for a major automotive company describing the airbag mechanism for experts (engineers), your description might look more like a set of specifications. Experts require less simple explanation, so a set of specifications will tell them what they need to know about the new technology.

Readers with a moderate level of specialization (e.g., colleagues) need more explanation than do experts. Moderate readers, though, have some background, so you would include for them the detail that you yourself would require for adequate understanding. Specifications may be too complex for the moderate reader, but a description of parts with visuals (a typical extended description) would be sufficient.

General readers or those with a low level of specialization require the most information in a format easily understandable. If you were to explain the airbag mechanism to consumers, for example, you would not emphasize the specifications used in their design and manufacture, but the ways in which the airbag mechanism is likely to work in actual practice. You would certainly include visuals and an explanation of the parts and how they work. Consumer advertising, brochures, and automotive how-to manuals target this consumer market.

WRITING THE EXTENDED DESCRIPTION

After evaluating the process of observing, recording, and naming details and considering the audience, we move now to the actual extended description you will write. Like the definition paper, the extended description is made up of a series of parts or elements that are important for your readership. Most of you will write for an audience with a moderate or low level of specialization, so choose a topic about which you know something and with which you feel comfortable. The extended description is made up of several parts: a *title* (that indicates the limited scope of the project), an *introduction* (with background material), a *body* (description of parts and their operation), and a *conclusion* (summary of parts and how they work together). Following the discussion of the parts of the extended description is a sample student-written description. Please refer to the student paper for examples.

Title

Earlier we talked about the importance of limiting the scope of the technical description so that the topic would be manageable. Make sure that your title reflects that limited scope. If you are describing just the blower assembly on a standard snow thrower, your title should reflect that (*Description of a Standard Blower Assembly of a Trademark Snow Thrower*) and not the whole machine (*Description of a Trademark Snow Thrower*).

Introduction (General Information)

In the introduction of the extended description paper, you will provide readers with a general introduction to your item. What is the item? What, in broad terms, does it look like? What, in general terms, is it used for or does it do? In addition, include general information in the introduction that might help facilitate reader understanding. Does the item have an unusual history? Can you offer general information about who is most likely to use it? Keep the introduction relatively brief and concise. Your audience is interested in a general overview of the item and the parts that make up the item.

Body (Naming and Explaining the Parts)

After the Introduction, you should move to a description of the parts that make up your item or object. As you did when you were recording detail, "blow apart" the subject in your mind. List the parts in the order you would follow, starting either from top to bottom or bottom to top. Provide a separate section (with a subheading) for each part and explain in as much detail as possible what the part looks like, how it is connected to the parts around it, and how it fits in the overall object or item you are describing. Depending on your topic, you may have only a few or many parts that require describing.

Conclusion (How the Parts Work Together)

After you have completed your description of each part and its relationship to other parts, you are ready to bring those parts back together for the reader. Indicate how the parts fit together to form the whole. After you have done this, explain to the reader how the whole object or item works—one complete cycle of operation.

Observe how one student wrote a technical description based on the standard automotive spark plug. Review the construction of the extended description, discuss the likely audience for the piece, and use the student paper as a possible model for your own work. Technical description is an important aspect of workplace writing that virtually anyone in any professional field may someday encounter. Understanding the component parts of the technical description will also make writing some of the other styles and forms present in this book a little easier.

Introduction, general description

DESCRIPTION OF A STANDARD SPARK PLUG

Introduction

A standard spark plug is a small metal and ceramic igniter that provides the spark needed to ignite the mixture of air and gasoline in the upper portion of a cylinder in an internal combustion engine, thus causing the explosive force that moves the piston within the cylinder to provide motive power to the crankshaft of an automobile, truck, or tractor. Spark plugs are also used in other types of stationary machinery where the identical explosive force may be used to power rotating gears as in sawmills, mowers, yard trimmers, leaf blowers, log splitters, and a variety of power winches. The spark plug operates on the principle of a confined explosion. The power created by ignition is able to travel in only one direction within a steel cylinder, thus increasing the force.

A standard spark plug consists of six main parts: metal terminal post, ceramic case, hexagonal metal base, threaded insert terminal, and two electrodes (see Figure 1).

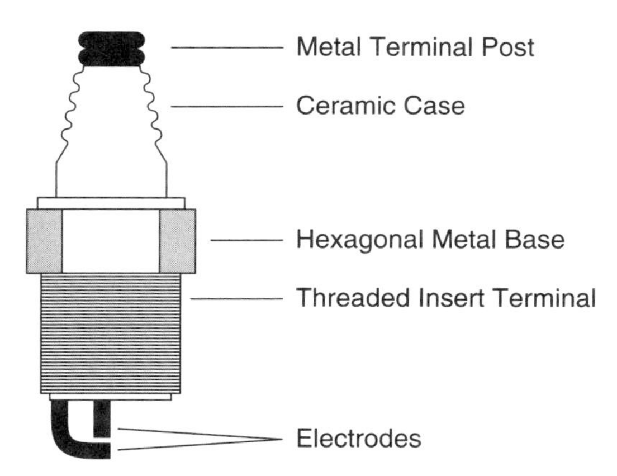

Figure 1. A Side View of a Standard Spark Plug

Description of parts and how they work

Description of Parts

Metal Terminal Post

The metal terminal post at the top of the spark plug receives an insert-connection from the end of a wire from the distributor, which has transmitted an electric spark from the coil,

continued

which in turn gets its initial charge from either a six- or twelve-volt battery. The terminal post is rounded, securely fixed to the ceramic case, and must be tightly fitted to the wire from the distributor.

Ceramic Case

The ceramic case is a pear-shaped cylinder that transmits the electric charge from the terminal post downward to the two electrodes at the bottom of the plug.

Hexagonal Metal Base

The hexagonal metal base, affixed to the bottom of the ceramic case, acts as the head of the metal bolt with threads that secure the spark plug to the cylinder head.

Threaded Insert Terminal

The threaded insert terminal is like the shaft of a metal bolt, except that in addition to its external threads, it is hollow to accommodate the two electrodes that protrude from its base. The terminal must be screwed down tightly to the cylinder head but not so tightly that the threads are scored (stripped). Spark plugs, therefore, should be tightened with a spark plug wrench.

Two Electrodes

The two electrodes that protrude from the bottom (see Figures 2 and 3) of the threaded insert terminal are of soft metal and are so placed as to emit a spark from one to the other. One electrode is straight and the other (L-shaped) is positioned from the side of the threaded insert terminal. The space between them, called the "gap," varies from engine to engine. The usual "gap" or "spark gap" between these two electrodes is thirty-thousandth (30/1000) of an inch.

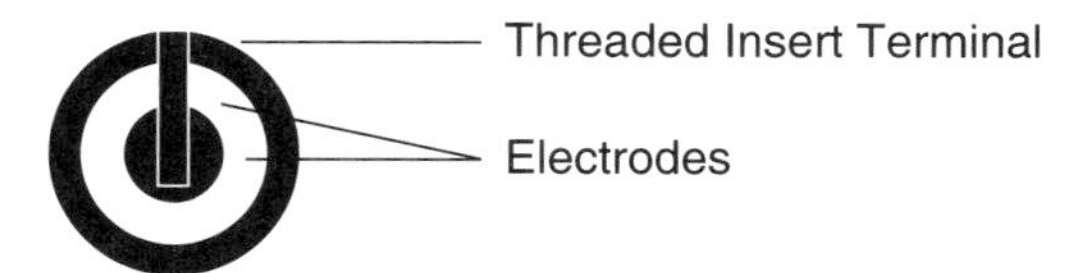

Figure 2. A Bottom View of the Spark Plug Base

continued

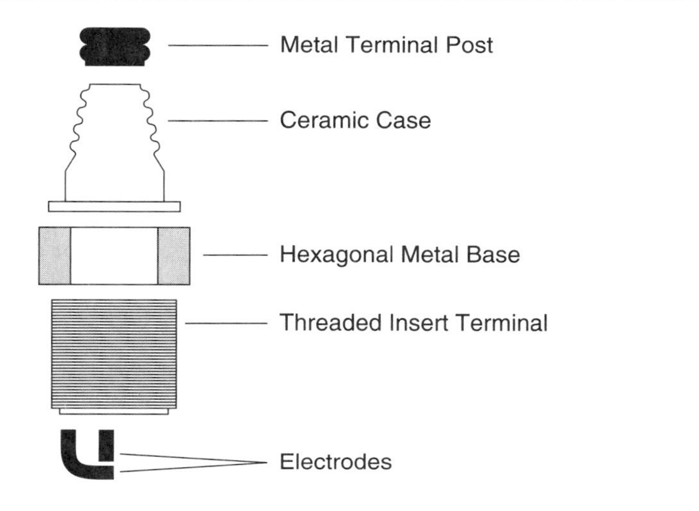

Figure 3. Exploded View of a Standard Spark Plug.

Conclusion and operating cycle

Conclusion and Operating Cycle

The standard spark plug ignites a mixture of air and gasoline in the automotive or machine cylinder, causing an explosion. The cylinder is vertical or slightly angled (as in V-8 engines), and the head of the piston is propelled downward in a closed block where its force is transmitted to a connecting rod moving the automobile or transmitting power from a stationary engine. The hot vapors from the internal combustion cylinder explosions are dissipated by timed exhaust valves. The cylinder confines the force of the explosion to move a steel piston head, harnessing that power to the purpose of the machine. The standard spark plug provides the vital spark for an internal combustion gasoline engine.

Exercises for Classroom Discussion

1. You have just been hired as a ______________ (fill in the blank with your job title or field of interest). Your supervisor has asked you to write a job description of your position, including all the responsibilities directly connected to the job and indirectly connected to the job. Your supervisor wants to include the job description you write in a report he has been asked to write on the personnel under his direct supervision. He may also use the job description for the purposes of future hiring. Your description, therefore, must be complete and concise.

2. You work as a ______________ (fill in the blank with your job title or field of interest). Your supervisor is going to hire someone to assist you in your work, but because that person will have a slightly lower level of specialization, he or she will need an orientation period to become accustomed to the job. As a result, select any item of equipment associated with your field and describe for your new assistant the equipment and how it is used in the job.

3. Select one of the following items, and in groups, write a description of the item and produce one visual that might accompany the item. Use either an overhead transparency or paper on which to depict your item.

- wheelbarrow
- syringe
- handgun
- compass
- distributor cap
- dot matrix printer
- water faucet assembly
- weaving loom
- potter's wheel
- socket wrench
- toaster oven
- atomizer
- seatbelt assembly
- bicycle braking assembly
- Bunsen burner
- carpenter's level
- stethoscope
- automotive jack
- fuel injector
- coffee maker

4. You are a pediatric nurse at a local hospital. Your supervisor is concerned that all new mothers and fathers be properly informed of the new laws mandating certified child restraint seats, so she asks you to prepare an extended definition on *child restraint seats,* the parts that make up the seat, where and how the seat should be positioned, and how the seat is to be used. You are quite familiar with the child restraint seats, but you never thought about how to write a description of one.

5. You are a water safety instructor for a midsized city in the upper Midwest. Your supervisor is very concerned about the number of personal watercraft on any given summer day in the lower harbor area of the city. Though the city has already made water safety classes mandatory for anyone operating personal watercraft, the instruction does not emphasize the importance of personal flotation devices (life jackets). As a result, your supervisor has decided that he wants you to incorporate a discussion of life jackets into all water safety instruction. He wants, as well, for you to write an extended description of the personal flotation device, its various parts, and how it is used for distribution to anyone who takes the water safety class.

SCENARIOS FOR TECHNICAL DESCRIPTION

The Pinging Modem

CONSIDERATIONS

In this situation, you are asked to look into an emerging technology to solve one of the company's communication problems. As you read this scenario, keep the following questions in mind:

- Who will be reading this document, and what will they be looking for in it?
- What research do you need to conduct to understand all the issues and technologies involved?
- What kind of information have you been given, and how reliable is it?

It's three o'clock in the afternoon at Motion Industrial, Inc. (MII), an industrial supply and consulting firm dealing in large, heavy machinery for manufacturing operations. You are quietly working at your desk, which is located in a little alcove off the hallway. As the newest employee, your workspace is temporary while your new office space is being outfitted with new furniture and wired for telephone and computer networks. One of the major advantages to working in the hallway, though, is that you can hear when the people near you are having problems with their computers. This is a good thing, because you are the new Technology Specialist for MII. You teach MII's two-hundred employees how to use word-processing, database,

Internet, and spreadsheet software programs through regular classroom-style and one-on-one sessions. You also work with MII's clients as a consultant whenever a particular piece of machinery requires integration into a company's computer network (some machines are computer controlled or have diagnostic devices to help pinpoint problems).

Today, your ears pick up telltale groans all down the aisle of cubicles nearest you—the Systems Analysis Department. MII's computers are programmed to dial automatically into an Internet server and check everyone's e-mail. The Systems Analysis computers perform this check at three o'clock.

"Server's down again!" you hear someone call. The phone rings, and you find yourself being summoned upstairs. You hurry to John Mackie's office. John, a Vice President at MII, handles all the technology acquisitions for the company and is your immediate supervisor.

"Just the person I want to see," John says, still staring at his computer. "I just called IS, and they say we don't have Internet access at all right now. Leroy said you had mentioned some ways of fixing this problem." John is referring to Leroy Smith, head of the Information Systems (IS) Department.

"There are other ways of setting up our systems. Right now, I'm getting calls and complaints from every department programmed to check their mail past one o'clock. Maybe we should set all systems to check mail before noon."

"I thought we took care of this when NetTech added more modems to their server," John says, referring to the company from which MII purchases its Internet access.

"They have a good number of modems," you reply. "I think the problem is that we're set up to hit the NetTech server twenty people at a time, and that's too much for them to handle with their existing equipment. I think we're also overloading our phone system at the same time–I've seen a few computers return errors because they couldn't get a dial tone."

"I don't like the idea of employees having to take care of their e-mail before noon," John says. "Some departments have morning meetings, customer contacts to get done . . . what else could we try?"

"Well, we could reprogram all the systems to dial in at staggered times," you respond.

"I talked to Leroy, and he's against that. He says the software we have on our network, the one that initiates the e-mail checks, can't do it in any fewer groupings than what we now have. We'd have to buy additional software to enable each computer to dial in individually. And that wouldn't solve my main problem with this whole Internet business."

"Which is?" you ask.

"Our access is way too slow. I don't want us paying for people to wait for sites to come up or for e-mail attachments to download. I received a file from a client, and it took ten minutes to save to my computer. Unacceptable." John frowns.

"Have you heard about cable modems?" you ask. "FutureCom is conducting a pilot project in our area," you say, referring to the local cable television company. "From what I've seen in their sales promotions, the speeds can be over one hundred times faster than our dial-up access."

"One hundred times faster?" John repeats. "Is it dedicated access, too?" You know that "dedicated access" means the company would purchase guaranteed access to an Internet service provider—essentially, renting out a certain number of the server's modems so that no one else could connect through them. "I think so," you answer. "And if they're using cable to provide the access, that means we wouldn't be using the phone system," you say.

"I could drop the phone lines dedicated to Internet access?" John says incredulously. "That would save us some money over what I was thinking."

"What were you considering?" you ask.

"Purchasing dedicated access from NetTech. It's very, very expensive and won't solve the speed problem," he answers.

You are not surprised—companies like NetTech often have ten or more customers for every available connection, banking on the theory that not every customer will dial in at the same time. To purchase dedicated access, MII would have to pay as much as NetTech would make by overselling their modems. "My problem with the cable modems is that they are so new, no one knows how they work or would integrate with our network," John continues.

You quickly realize that John is tactfully alluding to a problem you've already experienced at MII—resistance to changing equipment and procedures. Like most businesses, management often feel as though they've just taught everyone how to "work smarter" with their new computers, and then even newer software is needed and the learning curve becomes steep once again. You personally had a mild run-in with Leroy, a person you generally like, about upgrading the database software the Accounting Department uses to track customer accounts. While Leroy agreed that the accounting software was outdated, changing the software also meant purchasing additional equipment and programs for MII's computer network.

"Our network is fairly old," Leroy had said. "I spend most of my time just trying to keep everything functional." You also know that Leroy just received funds to replace the company's main hub, a device that hooks all the computers, printer, and file server together.

"Your technical documents to our clients have been excellent," John says. "I want you to find out how these modems work—how they hook up to our network—and write it up. I'll write a cover memo for this and call for a meeting with Leroy to go over the technical side. If hooking this up to the network won't cause major problems, then I'd be very interested in exploring how much it will cost us."

"I'll get right on it," you promise.

Later, back at your desk, you call FutureCom and are transferred to Tony Potts, a business sales representative for the new cable modem access.

"We're thinking about recommending your Internet access for our company," you say. "I need to know more about how the modems work and how they integrate into a local area network."

"Well," Tony says, "let's start with how our Internet access works. We'll start at our main office, where we have our direct connection to the Internet and our Internet servers. Now think about where you are—in an office, on a computer. When you want to access the Internet, you would simply open your Internet browser, which sends out a signal destined for the Internet. The signal then goes to your network and from there into one of our cable modems. The modem converts the signal so it will carry over coaxial cable—the same kind you use for your TV. From the coax, the signal goes to what we call a node, which is located in each neighborhood. The node converts the signal so it can run over fiber optic, which is what we have up on the telephone poles. The fiber runs into our main office, where it is directly connected to our Internet server, and thus the Internet as a whole. Reverse that process whenever you receive information from the Internet. This two-way capability is really what's new for us—we've been sending signals to people's homes for decades. Now they have the ability to send signals back down the line to us."

"That makes sense to me," you say. "How does the cable connect to our network?"

"Through the modem," he says. "Do you have a hub on your network?"

You answer in the affirmative. Tony tells you that one type of cable modem connects to the hub and would then allow everyone on MII's internal network to access the Internet.

"What about security?" you ask.

"You will need to take the same kind of precautions you take whenever you hook a network up to a device that accesses the Internet," Tony says and doesn't elaborate further. "Oh, did I tell you about how we make sure you're connected all the time?"

"No," you say. "Reliability *is* very important to us."

"When we put the cable modem in, we check how well it communicates with our main office through a process called pinging."

"Pinging?" you ask.

"Yes," he answers. "When the modem is connected and turned on, it sends a signal to the main office—like saying 'I'm here' to us. Our server automatically sends a signal back to the modem, which causes the modem

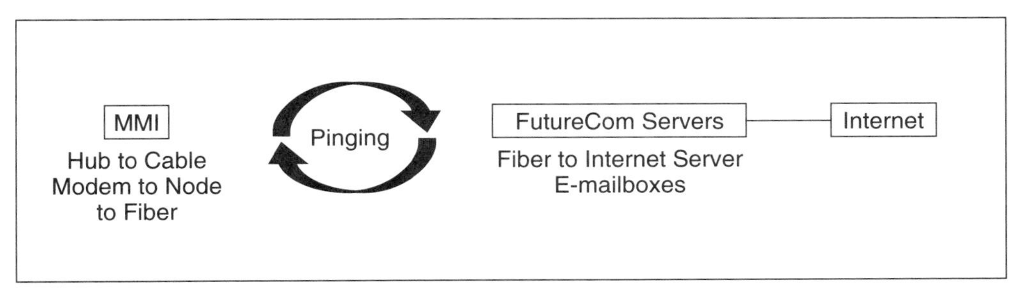

FIGURE 5.4 Cable Modem Signal Path

to send a signal to the main office, and so on. We can tell how long it takes your data to get from your offices to the Internet and back. Any slow downs or interruptions in the signal trigger a warning report that our network administrators see. It could be three in the morning and no one is in your office, but we know there's a problem that we probably could fix before you even get to work. It's a very unique system in that respect."

"Interesting," you say. You then ask about electronic mail services, which Tony assures you are part of the Internet access package. Tony offers to send you more information, but you realize it will deal with price more than the technical end of the service, so you ask him to send the materials directly to John.

After hanging up the phone, you reread the notes you'd taken during the conversation. You sketch a few things to help you organize the information you've been given (Figure 5.4).

While trying to get on the Internet and find more information about cable modems in order to verify FutureCom's claims about speed and reliability, you discover one major problem. Your Internet access is still unavailable.

Suggested Next Steps

- Review the guidelines for creating technical descriptions provided in this chapter.
- Refer to the appendix, Guide to Creating and Working with Visuals.
- Conduct additional research on cable modems–try technology-related magazines, journals, and the Internet.
- Create at least one visual that illustrates the features of cable modem technology.

Jack the Ripper

CONSIDERATIONS

In this situation, you are in a position to positively change the public's perception of your company, a forest products firm. As you read this scenario, keep the following questions in mind:

- Since, in this scenario, you will be creating a technical document on one unique piece of equipment, what kind of information will you need?
- Who are the people for whom you are writing?
- What are the political issues surrounding the creation and dissemination of your document?

You've recently begun work as a public relations officer for Terrace Forests, Inc., a lumber mill located in Washington state. Terrace Forests has been a major employer in Puyallup for almost one hundred years, although it was recently sold because of poor performance. Typically, you create press releases on company events (hiring, promotions, retirements, and profits) and handle or direct questions from the media. Your first more proactive project is to familiarize yourself with the company's most unique piece of equipment, using this information to interest various groups in touring the mill and thus generating company publicity.

Today, you are spending the morning with Dennis Ish, the Line 9 foreman. Terrace Forests operates twelve lines that produce various types and styles of lumber, from two-by-fours to particleboard to high-quality hardwood flooring. Line 9 is built around a specialized saw that produces construction-grade two-by-four, one-by-six, and four-by-four landscaping timbers, among other sizes of sawn logs.

Dennis greets you at the door to the Line 9 building and hands you a hard hat and safety glasses. "Need ear plugs?" he asks. You notice he is not wearing any, and you really want to hear his explanation of the sawing process, so you decline.

As soon as Bill opens the door to the saw room, the sharp wood sap smell and high-pitched whine of machinery overwhelm you. To your surprise, you notice very little sawdust in the air or on the floor of the mill. After asking Dennis how they accomplish this, he points to large vacuum cleaner–like hoses located over the saw blades. He yells in your ear so you can hear his explanation of how the sawdust is collected for other purposes. Tapping the standard brown clipboard in your hands, he says, "Sawdust and a little glue is all that is, really."

The main feature of the Line 9 building is the saw itself. You are interested in this particular piece of machinery, because you would like to bring forestry students from nearby State College to see this amazing saw. Dennis leads you to a small platform just a few feet from the blades, but divided by a thick Plexiglas protective barrier.

"May I present Jack the Ripper," he says with a big grin. You smile at the saw's nickname, recalling that its real label is the Dynablade 6-1000—certainly not as amusing. You take notes while Dennis continues to explain the machine.

The sawing unit is actually made of six circular saw blades, which simultaneously slice through the logs. A mechanized sliding arm in the saw table holds each blade in place and allows the saw operator to adjust the width of the lumber simply by programming in the desired board size. This means the operator can look at a saw log and decide what kind of cut will yield the best lumber and make appropriate adjustments without stopping the entire line to adjust the blades manually.

The vacuum system over the blades collects fine sawdust particles, but you notice larger chunks of wood dropping down below the blades and, more alarmingly, smacking against the Plexiglas barrier before falling below. Dennis explains that these pieces land on a conveyor belt below the blades and are used in making particleboard.

Another interesting mechanism is a large roller bar that holds the log down on its track. You assume it keeps the log from "kicking" up and jumping away from the blades. As you watch each log run through the saws; you see what Bill calls the "pusher"—a large hydraulic arm that pushes the log against the blades at a constant pressure. You roughly sketch Jack the Ripper (Figure 5.5 on page 132).

Back in your office, you check over your notes on the saw and decide to begin the documents. Dennis has informed you that only a few saws in the country can handle saw logs over seven feet in diameter, and even fewer multiple-blade saws can produce high-quality lumber from very large saw logs. Jack the Ripper is one of a handful of such saws in the country, you think.

Your boss and Terrace's Vice President of Operations, Roger Cassaway, knocks on your office door. "How did you like the Ripper?" he says.

"It was very impressive and very loud," you say. Roger has been pressuring you to dig yourself out of the typical press release jobs and come up with community-related projects that will yield good photo opportunities and press releases. You decide to tell him about your idea to bring in students from State College's forestry department.

"I'm sending them an information packet on touring, job shadowing, machinery studies, and internships," you say after catching him up on your

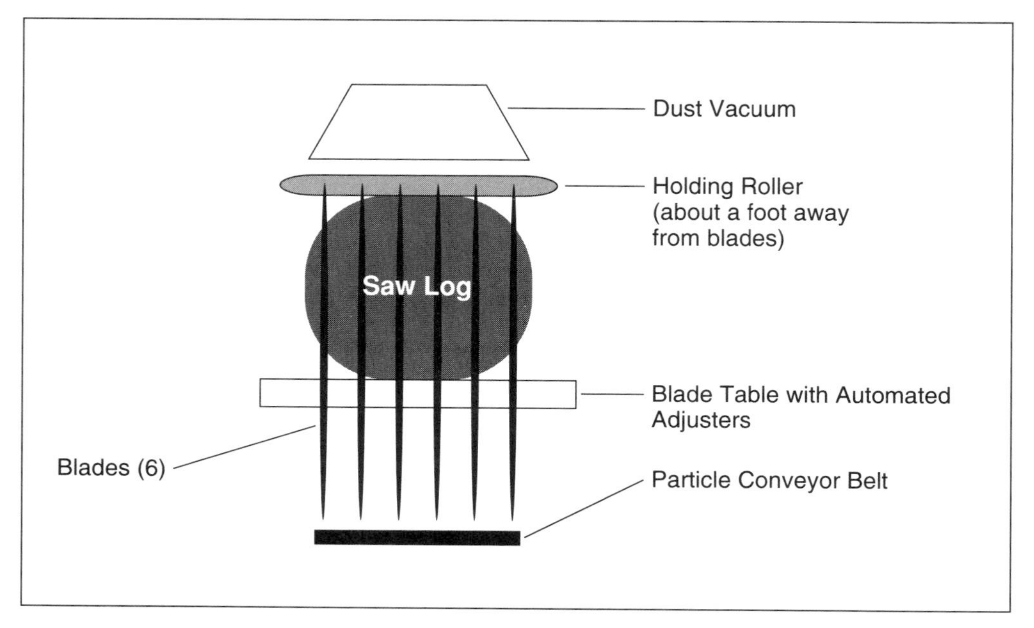

FIGURE 5.5 Jack the Ripper

ideas. "What I'm working on is a description of the Ripper for professors to use in their classes right now, instead of waiting until they schedule a visit. I want to give them something within the next few weeks, as a kind of favor to the department chair, Dr. Davies."

"Okay," Roger responds. "Sounds good to me. Write a cover letter to your paper on the saw, explicitly explaining why you've provided this information and that the college is under no obligation to pursue future collaborative projects. Make it a purely academic exchange—you provide classroom materials and the college investigates the possibility of doing some events with us."

"Sounds reasonable," you say. "I'll get right on it."

Suggested Next Steps

- Carefully review the guidelines for creating technical descriptions as provided in this chapter.
- Assemble the information from this scenario and create a technical description that includes at least one visual (refer to the Guide to Creating and Working with Visuals at the end of this text).

- Remember your own role and goals as a public relations officer. Be sure to include an explanation of the decisions you make in writing the cover letter and technical description in your Solution Defense.

Making Waves

Considerations

You are a part-time employee seeking to distinguish yourself in the kind of job you've always wanted. When asked to assist your boss, you jump at the chance. While reading through this situation, keep the following questions in mind:

- What technical skill levels will the document's readers posses?
- What are the problems associated with the wave-making device?
- How do these problems affect the creation of your document?

It took you two years, but you finally landed a summer job at the New City Aquarium. Your main responsibilities are helping to feed and clean up after the sea lions and otters and any other tasks required by your supervisors. While it has nothing to do with your studies in engineering, you enjoy working with the animals and the mental break the job provides.

Today, you report to work and are surprised to find a note taped to your locker:

> Please report to my office as soon as possible. We have an important project that requires your help.
>
> —J. Miller

The note's author is Jim Miller, a manager in the aquarium development projects division. You only met Jim once at a barbecue to welcome new staffers. You can only remember Jim as a very stocky man who looked more like a fisherman than a man who researches, plans, and helps fund new exhibits.

Jim greets you at his office door with a hearty handshake.

"What can I do for you?" you ask.

"I'm sure you're surprised to be called up here," Jim says. "But I remember you mentioning that you were studying engineering, and I have a project requiring that kind of work."

"I'm not an engineer yet," you explain.

"I know," Jim smiles. "I just need some help untangling what a real engineer worked up for our new shoreline birds exhibit."

"What's the problem?" you ask.

"It's these plans for the wave-making machine," Jim explains, handing you a manila folder. "I need to explain how it works for this grant I'm writing, and I 'm not sure I can figure it out myself."

"We're getting a wave-making machine?" you ask incredulously. New City's aquarium is relatively small, with a proportional budget. "Why do shoreline birds need waves?"

"Ah, engineers with no understanding of ecology," Jim jibes, but with good humor. "Shoreline birds feed off the small creatures that wash up on the beach. That's why it's so important to preserve specific types of beaches with the right mix of sand and rock—and the right wave height. We want to do more than just dump food at these birds. We need to demonstrate their dependency on the beach itself and the waves that feed them."

You nod appreciatively. "I have to explain how this machine is unique," Jim continues. "We've got to have a special kind of wave—not too big, not too small, and at the right intervals. And it can't make a smacking noise—that disturbs the birds. We had an engineer friend of mine develop it, but his specifications read like Greek to me. I need a complete description of the machine, but one that the people reading the grant will understand."

"I can work up a technical description for you," you say. "It may take me a little while, though."

"We have about a week," Jim says. "And I'll need you to do this in addition to your normal work. You'll get overtime pay for this project."

Your smile widens, and Jim laughs. "Get me a draft as soon as possible," he says and turns back to the work on his desk. You leave Jim's office, stuff the folder in your locker, and change quickly. The otters will be hungry soon.

At home, you open the folder and study its contents. You find a drawing of the wave machine (Figure 5.6).

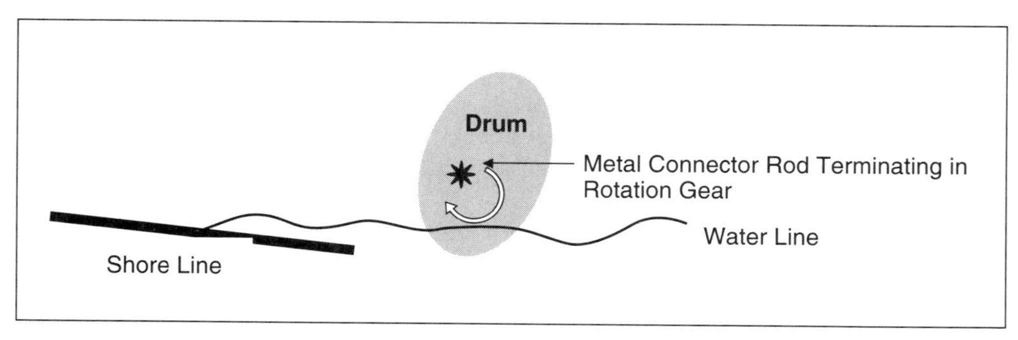

FIGURE 5.6 The Wave Machine

From what you can tell, the spindle arm rotates the oblong-shaped drum at a constant speed; however, the shape of the drum causes water to be displaced unevenly, creating the wave. The water that bounces back from the shore crests on the drum and is sent back as a smaller, gentler wave. The engineer estimated that five measurable waves occur with each rotation, with the first wave the largest and the rest decreasing in strength. A computer controls the rotation speed and thus the size and frequency of the waves. You see a copy of the engineer's report, which includes the following information:

> Mechanism produces volume distortion to the amount of 15 inches up a 10-degree "shore" slope. Succeeding distortions measured 9, 6, and 4 inches up the 10-degree slope. Some counter-wave action occurs every fourth rotation, with volume displacement measuring approximately 7 inches, with a 3, 2, and no measurable displacement until the next rotation. Some significant disturbance of sand under the drum occurred. Recommend using gravel under the drum and approximately 5 feet in front of and behind—should prevent excessive clouding of the water. Operation of the drum did not affect feed species population until water visibility dropped to near zero. NOTE: Heat from drum operation resulted in algae growth of unacceptable species (as determined by J. Miller). Problem not yet resolved; however, several low-heat prototypes are in the design phase.

A few days later, you catch up with Jim to ask him a few questions. "How much do you want me to include? I noticed that the engineer is having a problem with algae."

"Yeah," Jim says, "but he's close to solving that one, and we'll have it taken care of before we actually build the machine. Right now, we just need the money to start it up. The funders need to have confidence in this project."

You think you understand Jim's intent, but before you have a chance to talk further, he explains he is late for a meeting, asks you for a draft by tomorrow, and leaves you standing in the hallway with a sheaf of papers and unanswered questions. You realize you have no choice but to get a draft to him by the next day.

Suggested Next Steps

- Organize the information provided on the wave-making machine.
- After reviewing the guidelines for creating technical descriptions, decide how to represent your information within your document.

- Review the guidelines for dealing with ethical dilemmas (Chapter 3) and carefully consider how you will treat the design flaws. Discuss your choices in your Solution Defense.

The Glowing Globe

Considerations

In this situation, you are creating the technical documents for a new product, a decorative lamp. As you read this scenario, keep the following questions in mind:

- For what purpose will your document serve the company?
- For what other purposes might your document be used?
- What are the problems with this product, and how will you address them in your document?

You are a technical writer for Glowlife, a lighting fixture company specializing in novelty lamps. Glowlife's previous successes involve reproductions of older, Victorian-style lights. The newest prototype, and the focus of your morning meeting with a Glowlife project team, is a sixteen-inch-tall globe lamp on a cast-iron base. As the designers discuss what type of light bulb to ship with the lamp, you quickly sketch its shape (Figure 5.7). The glass globe will be manufactured in a variety of colors, from clear to dark

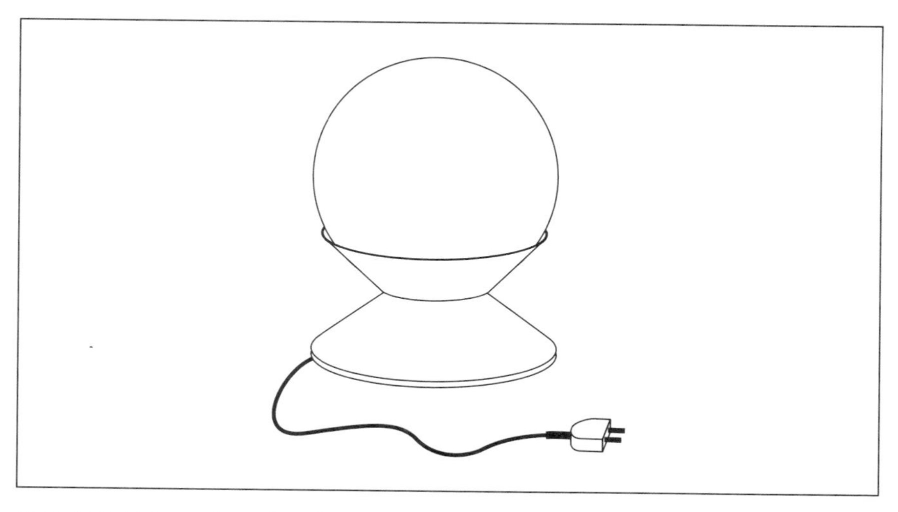

FIGURE 5.7 Sketch of the Globe Lamp

yellow to a patterned green, blue, and white earthlike treatment. The base is of two-part, cast-iron construction, with four small screws to hold the two pieces together.

"Will the base be disassembled in the box?" you ask Kevin Maxin, Glowlife's chief packaging designer.

"No," answers Tracy Sherwood, the marketing manager assigned to the globe lamp project. "We don't want customers to have to do anything more than put the glass shade on and plug it in."

Kevin looks at Tracy, then at you, and shrugs. "I can make the package any way you like it," he says.

"How does the glass attach to the base?" you ask.

"Gravity," answers Mickey O'Shay, the lamp's designer. "The pedestal is deep enough that the globe simply rests in it. I've designed it so the opening in the globe is just big enough for the light bulb and socket. It's actually pretty hard to knock it off the base." As she explains this to you, Mickey moves the lamp over the table with quick, jerky motions. You hear glass rattling against metal, but the globe does not visibly tip out of the base.

"What kind of socket do we have in this one?" you ask.

"A standard 40 watt," Mickey answers.

"What did you decide on for the switch?" Tracy asks.

"The top portion of the base is brass plated instead of cast iron," Mickey says, "so we can make it a touch lamp. Touch the top part of the base, and the lamp turns on. Touch it again, and it turns off. This eliminates the need to use a standard switch on the cord."

"Our marketing analysis indicates that touch lamps will sell much better," Tracy adds. "I'm pleased you went with it," she says to Mickey.

"And I'm surprised," Kevin says. "I wasn't really paying close attention, because it didn't affect me, but weren't you worried about heat from the lamp making the top portion too warm?"

"I added heat insulation around the socket," Mickey says. "As long as the user only puts 40-watt bulbs in the socket, it should be fine."

"I'll be sure to add that in the description," you say. "Speaking of warnings, how warm does the glass lampshade get?"

"Not very warm at all," Mickey replies. "Not enough to start a fire. That's one great thing about globe lamps. There's no exposed bulb."

You decide you have enough to begin your description of the lamp. Every time Glowlife begins manufacturing a new product, someone must write a complete technical description, which then serves as the basis for in-package product literature and marketing materials. The meeting breaks up, and you gather your papers to leave. That's when you notice several scratches on the table where Mickey had been moving the lamp back and forth. Curious, you pull on the lamp cord to move the fixture closer. To your surprise, you see that the base of the lamp makes four distinct scratch marks. You remove the globe from the fixture and turn it over. Puzzled because you do not readily see anything that should cause scratches, you run your finger around the ring of metal that makes up the circular base. That's when you feel a few burrs, or pieces of metal that occur when the mold used to cast the metal isn't perfectly smooth.

"Mickey," you say loudly, just as she's leaving the meeting room. "Will you look at this?" You point to the scratches on the table and show her the burrs.

"I know," Mickey says. "Unfortunately, we've already made thousands of these base pieces. I've asked for better molds, but they're so expensive to make that I doubt we'll get new ones before the second round of production."

"What about putting felt on the bottom?" you ask. "

"Wouldn't work," Mickey replies. "The base is too thin, so felt wouldn't really stick to it."

"I see," you say. You finish inspecting the base and return to your office. *I should indicate that the base might cause scratches on wood or other soft surfaces,* you think, but you are hesitant to do so. For another lamp project, the marketing department (not Tracy, but another marketing assistant) was very upset at your insistence upon warnings for a lamp that put out enough heat to start a fire on a typical blanket. All Glowlife products are safety-tested in a variety of different scenarios, from standard electrical tests to performance under abnormal situations, such as having a blanket, facial tissue, or pillow thrown over the fixture. While the globe lamp is fairly benign in this respect, you realize that consumers could complain about the lamp scratching their tables, and many might return the lamps, hurting sales and generating negative publicity.

You sit at your computer, getting ready to create a draft, even though you are not entirely sure how you will handle this situation.

Suggested Next Steps

- Review the guidelines for creating technical descriptions as outlined in this chapter.
- Decide what information you will include in your description, and how you will visually represent the product. Discuss your decisions in your Solution Defense.
- Review the guidelines for dealing with ethical dilemmas (Chapter 3) when considering how to handle the scratching problem. Justify your decisions in your Solution Defense.

Soil Incineration: Technical Description Troubles

CONSIDERATIONS

In this situation, you will need the technical data supplied by the soil incineration scenarios from the previous chapters. As you read about these new developments in the project, keep the following questions in mind:

- What is the function of the incinerator's baghouse?
- What are the political issues surrounding this phase of the project?
- What are your personal stakes in the success or failure of the incineration facility?

After surviving the creation of the technical definition for the soil incineration project, you are almost looking forward to completing the technical description. As you review the other documents you've created and gathered on the project, you make a mental note to improve the visual you used for the definition and mark information you will need for the new document.

The incineration project is still the hot topic of debate in your community. The Protect Our Land (POL) organization has taken to stuffing your mailbox, from time to time, with anti-incineration fliers, and you hardly ever go out for lunch anymore—no one will talk to you, and you've grown tired of the angry stares from some of the more forward residents.

You review the specifications for different parts of the incinerator, making the following notes:

For the low temperature, thermal treatment, drier drum—volatile organic compounds (VOCs) must be a maximum of 25 percent of soil weight. Virgin soil can be added to dilute VOCs in soils testing 26 percent or higher by weight.

→ Drier drum, under ideal moisture and soil grain (minimal soil clumping) conditions will decrease VOCs to 10 parts per million (PPM). Gas emissions for soil under 25 percent VOC will meet state standards. Manufacturer does not guarantee gas emission, efficiency, or production rates above 25% VOC.

→ Cyclone mechanism—separates particulate matter from gas stream (coming out of drier drum). Good antipollutant for gas emissions—can decrease VOC emissions by up to 15 percent but manufacturer recommends secondary devices (baghouse) to remove finer particles.

continued

→ Baghouse dust collection system—considered best defense against VOC particles becoming airborne (could travel long distances).

After finishing your notes, you decide to take a break and get some coffee. You find Jack in the break room, waiting for the coffeepot to finish brewing. You exchange information on a newer Donnelly Engineering undertaking—the construction of a bike and hiking trail through a particularly sensitive lakeshore ecosystem. Jack asks you to follow him back to his office to discuss the incineration project.

"There's been a new development," Jack begins. "I just spoke with Keith Robbins, our contact with the firm financing the incinerator's construction."

"By the way," you ask. "Why aren't we including the name of the company that's financing the project and the management company in any of the documentation? Wouldn't the county want to know who would be responsible on a daily basis? I think some of our problem with negative publicity is all the secrecy."

Jack opens his desk drawer and pulls out a sheaf of papers. You recognize them as the same POL leaflets and flyers you find in your mailbox. "Our job is to protect the financiers and operators from this kind of harassment," Jack says. "If Donnelly's name isn't enough to alleviate their concerns, then I don't know what could."

"Understood," you say. "What's the new development?"

"We might not have the start-up funds to put the baghouse in until after the first year," Jack answers. "I've read the specifications for a drier drum, afterburner, and cyclone particulate removal system, and I know that we can get in under the emission regulations."

"But the manufacturer doesn't recommend not having a baghouse," you say. "And we'd have to make sure that anything going into the plant is at most 25 percent VOC—that means adding quite a bit more clean soil than we originally planned."

"Regardless, you need to justify the possibility of not having a baghouse. We don't need to give POL any more ammunition than they think they have already," Jack says.

Back in your office, you think over the situation. A few days ago, you heard Jack literally yelling in his office, and later heard that he became very angry upon learning that POL had convinced the county zoning commission to conduct further research on soil incineration projects in other communities before granting a rezoning (from farmland to industrial). Donnelly had been counting on Jack's close ties with the commission to prevent such delays, but it obviously hadn't worked.

Your own job may very well depend on how well you can convince the people reading your description that the incinerator will not cause the rampant environmental damage POL claims. Just last week, Kyle Donnelly (the company president) stopped by your office to congratulate you personally on an unrelated project. It didn't take long for Mr. Donnelly to ask about the incineration project's status. From his level of familiarity with the project, you realized that he had read the documents you've already produced.

Suggested Next Steps

- Carefully review the guidelines for creating technical descriptions provided in this chapter.
- Compile the information you need to write a technical description of the incineration plant.
- Review the guidelines for dealing with ethical dilemmas (Chapter 3). Be sure to explain your decisions concerning how you dealt with the change in plans and company pressures.
- Be sure to include at least one visual in your document.

CHAPTER 6

Technical Instructions

In previous chapters we've discussed the relative difficulty of writing technical definitions and descriptions for different audiences. In this chapter we'll evaluate *technical instructions,* a very important form that is a large part of the kind of assignments technical writers most often get in the workplace. Think of technical instructions as explanations to your audience on how to do something. You might explain the relatively simple steps necessary in securing a bank loan or something more complex like how to plot a course on a ship's radar system. Regardless of the topic, technical instructions usually involve a series of steps or guidelines that will provide the audience enough information to duplicate the process. Examples of technical instructions for a general audience are:

- instructions for programming and using a video cassette recorder (VCR)
- instructions for assembling a child's swing set
- directions for installing a home security system
- directions for installing a new washing machine

If you have read the instructions for filing your taxes or if you have used the instructions for assembling a small appliance, you have experienced this form first hand. While the instructions, in some cases, may seem simple on the surface, writing a good, usable set of instructions involves some planning and, most importantly, consideration of the audience.

PLANNING FOR TECHNICAL INSTRUCTIONS

Before you even think about writing a set of instructions, consider the following important points first. After you've read through them and the section on audience analysis, study the set of guidelines for writing your paper. Read all of this material before drafting the paper. These important points will help you think ahead to the kinds of information your audience, regardless of technical expertise, will find most useful.

1. To write the best process analysis possible, select a topic that you know well and that is, if possible, connected to your major. If you are a nursing student, perhaps "how to take a patient's blood pressure" would be a good choice. If you are an accounting major, a paper on "how to balance an account" would work well. Regardless of your choice, avoid topics with which you are unfamiliar. Remember—your audience is relying on you to explain a theoretically unknown process.
2. Narrow the topic to something manageable in three or four pages. Instead of explaining "how to build a house," consider explaining "how to install a bay window in a new house." Remember—process analysis involves careful discussion of each step in a set of instructions.
3. Your audience will appreciate visual aids, so plan to have at least one or two visuals to accompany the text of your paper. If you've ever attempted to assemble anything involving small parts (e.g., a child's dollhouse), you know how valuable the visuals are for depicting how parts fit together. Remember—visuals are the one language we all "speak," so use visuals particularly when you are explaining something complex.
4. Make a list of all the equipment or parts a reader might find necessary to duplicate the process. You will, in all likelihood, include that list in the paper you write, so consider all the things a reader would need. Does the process require specific tools (e.g., a Phillip's screwdriver) or general equipment (a hammer and nails)? Remember—do not omit important equipment from the process analysis. If you are explaining how to gap spark plugs manually, your audience must have a feeler gauge.

Audience Considerations

As you've certainly learned by now, analyzing your audience is important if you are to write instructions that can be used effectively. Have you ever had difficulty programming a VCR to record a television show? Have you ever given up on your computer manual and asked a friend for instructions instead? If you have ever faced either of these scenarios, then you know firsthand the frustration that can accompany instructions that often were not written with the needs of the audience in mind.

In writing technical instructions, the needs of the audience are of paramount importance because undertaking a process can be—depending on the process itself—risky or even dangerous. Thus, careful, step-by-step instructions may be written for someone with an expert level of technical background, a moderate level of technical background, or a low level of technical background. Consider, for example, that you were asked to write a set of instructions for properly felling a tree. The instructions for an expert logger would certainly be different than those written for a novice logger felling a tree for the first time. Think, therefore, very carefully the needs of your audience before you begin drafting the process analysis paper.

Because most of you will probably choose to write technical instructions for readers with a low level of technical background, consider some of the following points. Readers with a low level of technical expertise often benefit from

- *Instructions written in the active voice and imperative mood.* Rather than suggest to readers that they ought to do something, *tell them* what needs to be done. Consider the following examples. Which is in the active voice and imperative mood?

  Example A. In a marine engine, it is advisable to start the engine only after the blower has been operated for roughly ten minutes.

  Example B. Start a marine engine *only after running the blower for ten minutes.*

  In Example B the writer began the instruction with the verb first (active voice), a good idea when writing technical instructions. Also, notice that in Example A the writer suggested the reader do something, while in Example B the writer insisted that something be done (the imperative mood).
- *Steps in chronological order.* Decide in which order you will present your technical instructions and remain consistent. If you are explaining to readers how to change a tire, will you begin with the removal of the flat tire? Will you begin by explaining how to jack up the car? Will you explain only how to put on the new tire? Decide on a plan and follow a logical order so that readers are not confused.
- *Complete instructions.* If you are explaining how to change a tire and you tell your readers to first hand-tighten the lug nuts on the spare tire, explain why the nuts should be first hand-tightened. If you go on to tell readers to finish tightening the nuts with a wrench, explain what kind of wrench they will require.

- *Headings.* Use headings and subheadings to move readers from one major set of instructions to another. For example, if you choose to explain how to change a tire, how will you group the steps into workable segments? You might have three major headings: *Jacking up the Car, Removing the Flat,* and *Putting on the Spare.* Within each of those major segments, you plan to present a series of steps. The headings help your readers move from one large task to another.

All of these points are important in writing the technical instructions paper. The following sections will guide you in writing through the three key areas of the paper itself: the overview and introduction, the instructions, and the conclusion. Each section of the technical instructions, when well-written, plays a vital role in assisting readers undertaking the process.

Overview and Introduction

The overview and introduction prepares readers for the instructions and explains what tools and safety concerns readers need to think about before beginning the process itself.

Preparing for the Process

Use the introduction of your technical instructions paper to explain to readers why the process is important or meaningful. If you are explaining how to change a tire, note in the introduction that knowing the process can make a difference when a motorist is stranded on the side of the road. If you are explaining how to perform cardiopulmonary resuscitation (CPR), mention in the introduction that learning the steps to CPR can save someone's life.

In general, use the introduction to set up the process, explain its importance, and list the tools required to perform the process and the safety measures necessary to insure a safe process.

Tools and Safety Concerns

To perform a process properly, readers need an accurate list of tools or implements necessary to complete the job. For example, if you are explaining how to construct a birdhouse, readers need to gather the necessary equipment: hammer, nails, saw, level, drill, pine boards, and any other materials you believe to be important to the process. List the tools or implements early in the paper so that readers are prepared to begin when you move to the instructions.

In some cases, you may explain a process that requires special care or caution. If you are describing how to weld a steel seam, readers must take special care because of the heat and potential eye damage associated with the acetylene torch used in the process. When the overall process requires constant care and attention to detail, notify your readers early in the paper using the word **NOTE.**

> **NOTE:** Use care throughout the welding process. Always wear eye protection to prevent injury to the eyes from the acetylene torch.

If you need to caution or alert readers later in the process, do so just prior to the step in question. Use the word **NOTE** to indicate special consideration, **CAUTION** to indicate potential physical harm or **WARNING** to indicate the potential for harm or even death. Do not, however, overuse cautionary notes or warnings. Readers may have a tendency to disregard such notes and warnings if you include too many.

Instructions

The body portion of the technical instructions paper is made up of step-by-step instructions necessary to complete the process. Because the instructions are in many ways the most important part of the paper—the process itself—spend extra time and attention on the presentation of the steps. You've already learned some points for increasing the readers' understanding of the process, but consider the following specific tips for the instructions.

- Present each step with the detail necessary to complete the step. If you are explaining how to bake bread and you tell readers to "let the bread rise," you've neglected to explain how to prepare the bowl for the bread, where to put the bread to rise, whether or not to cover the bread, and so on.
- As you write the instructions, use words that signal changes—*transitional* words or expressions. Words like *next, then, after, before,* and other such words help readers to follow the shifts from step to step in the process.
- Use visual aids to help readers understand the process. Visuals are particularly useful when you are explaining a complex step. Notice the visuals used in the student-written paper at the conclusion of this chapter. Refer to the appendix, Guide to Creating and Working with Visuals, for more information on using visuals or graphic aids.

- Use numbered steps when you are explaining how to complete a specific sequence. If you are explaining how to type blood, use numbered steps when you describe drawing the blood.

> First, using a lancet, prick the subject's index finger until a bead of blood rises from the fingertip. Second, collect the blood sample by placing a small pipette just under the blood bead. Third, place an alcohol-soaked cotton ball on the subject's fingertip to slow the flow of blood. You are now ready to begin the process of typing the blood.

All these points are useful in helping you put together the technical instructions analysis paper. Refer to the student-written paper, Instruction for Making an Eye Splice in Rope on pages 149–153, as an example of the instructional portion of the paper. Do not, though, leave out the important conclusion of your process paper.

Conclusion

The conclusion is your opportunity to summarize the process itself and remind readers of the relative importance of the process. Some writers like to use the conclusion as a place to suggest variations on the process. For example, if in your technical instructions paper you explain how to change the oil in a vehicle, you might suggest to readers that changing the transmission or radiator fluids are somewhat similar, money-saving processes that they may wish to learn at a later date.

TIP

Some writers use the conclusion of the process analysis to provide the reader trouble-shooting tips. If, for example, your technical instructions paper explains the procedure for installing an automatic igniter in a gas grill, you could use the conclusion of the paper to explain what might have gone wrong if readers have difficulty getting the grill to light using the automatic igniter.

The following student-written technical instructions paper demonstrates how the paper should be put together. Whether your instructor permits you to choose a topic or assigns one of the up-coming scenarios for technical instructions, one of the most important points in any technical writing project is consideration of the audience. Carefully consider the needs of your audience before you begin writing the paper.

Introduction and overview

INSTRUCTIONS FOR MAKING AN EYE SPLICE IN ROPE

Introduction

An eye splice in rope consists of a permanent loop or "eye" at the end of the rope (or "line") so interwoven with the regular strand of the rope that it will not pull apart, even under great strain or tension. Eye splices serve a variety of uses, from mooring lines, anchor ropes, towing lines, and sliding loops aboard ships and boats, to many uses in industry. Eye splices might also be used even in such sports as mountain climbing, camping, and backpacking. Splicing is an ancient and useful skill.

> **WARNING:** An eye splice will be as strong as the parent rope only if you perform the following instructions properly. A loose or sloppy splice could result in accident or injury. If you are performing a splice for the first time, practice on several rope remnants before attempting a splice for serious use.

To make an eye splice in a conventional three-strand rope, you will need the following equipment:

a marlinspike or fid

a spool of waxed twine

a roll of electrician's tape

A *marlinspike* is a traditional sailor's tool consisting of a forged steel rod, usually about six inches long, rounded on one end and sharply pointed on the other. A *fid* is the wooden equivalent of the marlinspike, although usually a bit larger. Use either tool to create strong eye splices.

These instructions are for use with conventional manila or hemp rope only. They may or may not be effective in splicing rope of synthetics, such as nylon or dacron, depending on the interweave and number of strands within such rope. The major steps in creating an eye splice are

1. unravelling the three separate strands
2. wrapping a small piece of electrician's tape around each strand to prevent it from unravelling further

continued

3. making the splice
4. smoothing the splice

Required steps

Instructions

1. Unraveling the Three Separate Strands. Simply take the end of the rope (called the "bitter end" by seamen) and unwind them into three separate strands. The length of the strands determines the size of the finished eye splice or loop. The larger the finished loop, the longer must be the beginning strands. See Figure 1.

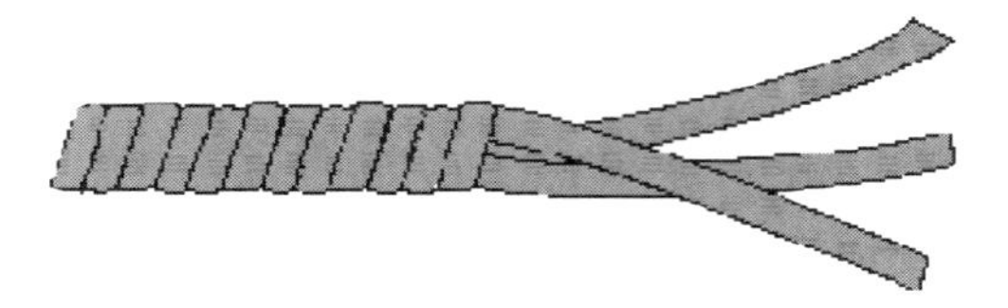

Figure 1 Unraveling Three Separate Strands

2. Wrapping the End of Each Strand. Carefully wrap a small piece of electrician's tape around the end of each strand to make sure the rope tip will not unravel. Wrapping also makes it easier to push the end of the strands through the weave of the parent rope. This procedure is called *whipping.* See Figure 2.

Figure 2 Wrapping Strand Ends

3. Making the Splice. For convenience, mentally number the three separate strands, the middle strand (number 1), the left strand (number 2), and the right strand (number 3). Determine where the bottom of your eye splice (loop) will be, and with your hands, twist on the parent rope to make a small opening. Next, slightly enlarge this opening with the

continued

marlinspike or fid. Now pull strand number 1 through this opening all the way to the untwisted body of the parent rope. See Figure 3. Continue this procedure, but now tuck strand number 2 under the lay (coil) of the parent rope immediately to the *left* of number 1.

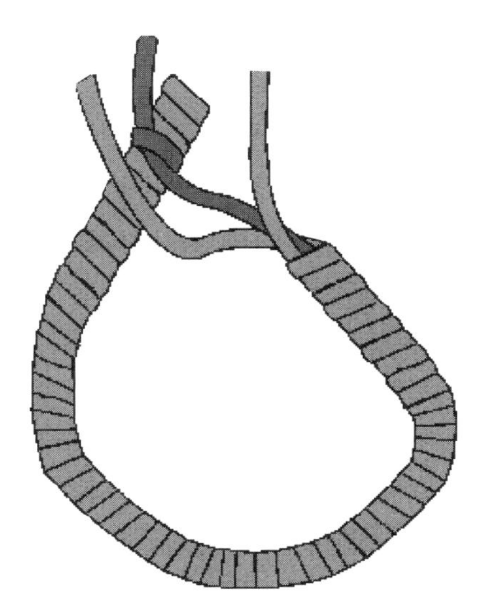

Figure 3 Making the Splice

> **NOTE:** Never tuck two separate strands under the same lay (coil) of the parent rope. Each strand must be inserted under a separate loop made by the marlinspike or fid. See Figure 4.

Follow the procedure above, but next tuck strand number 3 in its own parent loop to the *right* of number 1. See Figure 5.

You have now formed a loop (beginning eye splice) in a length of rope. Simply continue this process, creating new openings in the parent line and inserting the appropriate strand through its own loop. Pull each strand tightly against the the parent rope. Be sure to tuck each strand in every other loop, never in the same loop as another strand. Make three or four more tucks, depending on the length of the strands and the eventual size of the eye splice.

continued

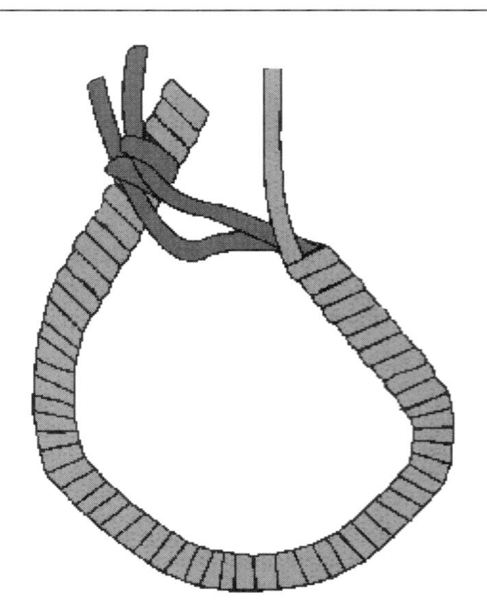

Figure 4 Inserting Strands under Separate Loops

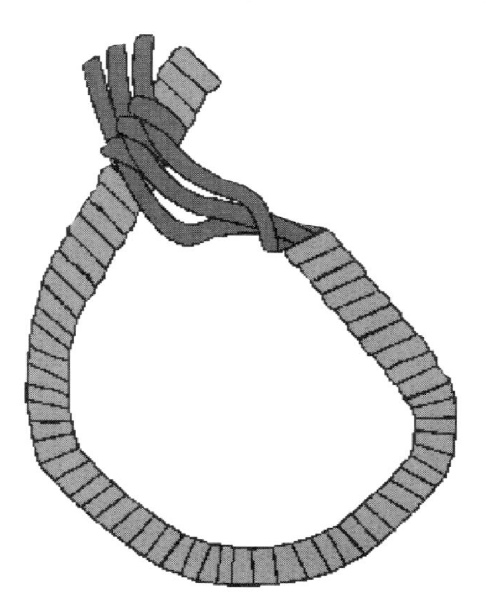

Figure 5 Forming the Loop

4. Smoothing the Splice. The ends of each strand will be protruding from the parent rope at the end of the splice. Smooth these ends as much as possible into the parent rope.

continued

Do this by *rolling* the completed splice in the palm of your hands (for small rope or line) or rolling it under your foot on a smooth, hard surface (for large or unwieldy rope or line). See Figure 6.

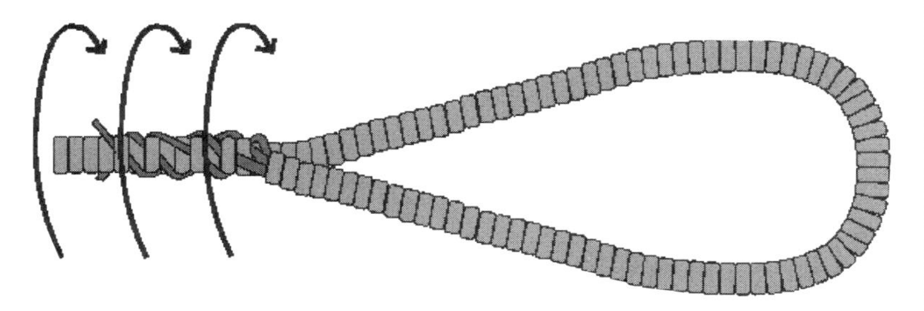

Figure 6 Smoothing (Rolling) the Splice

Complete the process of "smoothing" the splice by winding (whipping) waxed twine around the ends of the tucked strands. Such whipping can be of crucial importance if the eye splice is to be pulled through a block and tackle pulley or over a balk of timber, steel shaft, or other protrusion. See Figure 7.

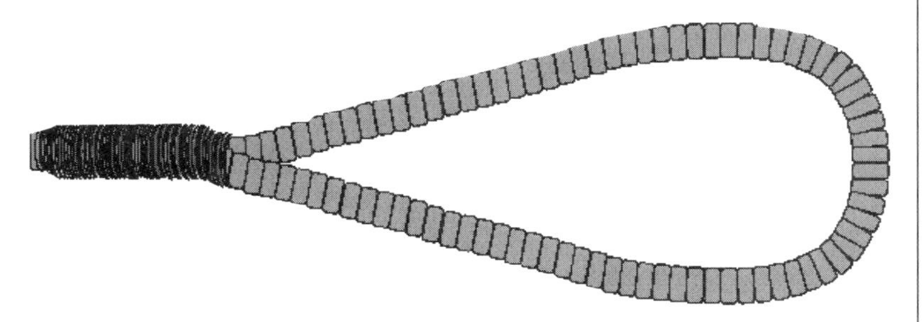

Figure 7 Completing the Splice

Conclusion

Conclusion

The uses for a properly constructed eye splice are virtually endless, limited only to the needs and imagination of its users. By following these instructions, you will fashion a permanent and very strong loop on the end of a straight rope (line). With practice and patience, you'll develop a valuable and versatile skill.

Exercises for Classroom Discussion

1. You work as an intern in the registration and scheduling office at your university. Your direct supervisor, the Assistant Registrar, recently approached you and asked you to write a "student-oriented" set of instructions on how to register for classes. She asks you to include any "inside" information you may have, such as how to gain enrollment in courses that are traditionally always filled early, the best places to register without long lines, and so on. The instructions you write will be sent to students with the next semester's course information materials.

2. You work as a ____________ (fill in the blank with your job title or field of interest). Your supervisor is going to hire someone to assist you in your work, but because that person will have a slightly lower level of specialization than you have, he or she will need an orientation period to become accustomed to the job. As a result, you have been asked to select any item of equipment closely associated with your work and explain to your new assistant how it is used in the job he or she will hold.

3. Select one of the following basic processes, and, in groups, write a brief process analysis and produce one visual to accompany the item. Use either an overhead transparency or paper on which to depict your process.

- how to make coffee
- how to plant a tree
- how to shoot a target
- how to use a compass
- how to check the oil level in your car
- how to use a computer printer
- how to wash your car
- how to pick a Christmas tree
- how to make a clay pot
- how to jack up a car
- how to balance a checkbook
- how to play ____________ (fill in any game)
- how to pack for a trip
- how to replace a bicycle tire
- how to chop wood
- how to make up a bed
- how to check blood pressure

4. You are a nurse at a local pediatrics hospital. Your supervisor asks you to prepare for new mothers a two-page handout explaining the basic steps of bathing, bottle feeding, and clothing a newborn. You are quite familiar with all three processes and you expect the job to be pretty easy. Once you sit down to write it out, though, you are surprised at how difficult it is to explain to someone else a process that you have found relatively simple.

5. You are a water safety instructor for a midsized city in the upper Midwest. Your supervisor recently informed you of a program the city has instituted that uses high school students as assistant lifeguards at the two local pools. These students have had some safety training, but none of them know how to perform CPR. The city will not authorize them to work, even as assistants, without such training. You must write up the steps for performing CPR and prepare a handout that the students will use as a reference after they complete the CPR course.

SCENARIOS FOR TECHNICAL INSTRUCTIONS

Finding the Money

Considerations

In this situation, the bulk of your work is researching the financial aid application process. As you read the scenario, keep the following questions in mind:

- What do you know about obtaining financial aid at your college or university?
- Who are your readers, and what are they looking for from your information?
- How much detail is required to make this information useful to students at other schools?

You landed a summer internship with a small publishing company, Matheson & Matheson (or "M&M" as your co-workers call the company). M&M specializes in tabloid publications, mostly for the Seattle club scene. You're job has been to rotate from one publication staff to another, getting as much exposure to the different aspects of the company as possible in the few months you have before you return to college. So far, you like the job, especially the relaxed office atmosphere.

Today, you are working with the staff of *Campus Matters,* a monthly tabloid for the students of the Seattle area's several universities and community colleges. So far, you've tagged along on interviews with a few local bar bands, and you've helped edit a couple of feature articles—one of which, titled "Living on a Shoestring Budget," you found very interesting, because your internship doesn't pay all that much.

You are performing some of the more menial morning tasks—getting water for the coffee pot, dropping mail off at the appropriate cubicles—when Kara, the editor of *Campus Matters,* calls you into her office. You ask her to wait a second so you can get the coffee going (you know better than to keep the caffeine-addicted writers waiting). You return in a moment and take a seat, wondering what she has in mind for you today.

"I vaguely remember you mentioning that you had some experience in technical writing," she says.

"A little," you answer. "Just the basic college course."

"Good enough," Kara says, smiling. "Myself, I only took poetry classes, and look how much I get to use it." You smile in return.

"We want to do an informative article for the next issue," Kara says. "And I think we've decided on a topic. The article will be about applying for financial aid, and we want you to do it."

"Great," you say. "What kind of financial aid?"

"What kind?" Kara says. "There are different kinds now?"

"Well, yes," you say. "I mean, at most colleges, you fill out the basic form, and then they tell you what kinds you can get. Then you apply for those things, like student loans or work–study programs. If you qualify for government grants, they just give them to you, most of the time."

"Why don't you start at the beginning of the process," Kara says.

"Sure," you answer. "When do you want this?"

"As soon as possible," Kara answers. "We want to do a special layout on it—something that will make it look fun."

"Financial aid applications are anything but fun," you say, speaking from experience.

"Try to make it sound as painless as possible," Kara says, smiling. "That's all. Get a draft on my desk as soon as you can."

You leave her office and return to your cubicle. *The first thing,* you think, *is to call someone at the university who can help me.*

Suggested Next Steps

- Review the guidelines for creating processes as outlined in this chapter
- If you are unsure about the intricacies of applying for financial aid, contact the appropriate office at your college and request information. Indicate the sources of your information in your Solution Defense.
- What kinds of information will students need to know before applying for financial aid? What expectations should they have during and after the process? Discuss how you handle these issues in your Solution Defense.

- How do you handle the tone and stylistic issues, given that this piece is for a magazine article and is, therefore, not *just* a technical document? Do you try to be humorous, or do you write a "straight" piece? Examine your choices and defend your decision in your Solution Defense.

Cellulose Insulation Installation

CONSIDERATIONS

In this scenario, you will receive information needed to write instructions; however, the bulk of that information is given to you verbally. Keep in mind the following questions:

- What pieces of information are necessary to the mechanics of the process?
- What information provides general understanding and helps avoid problems?
- What are the political issues and legal ramifications surrounding the writing of these instructions?

As a technical writer for New Age Construction Materials, Inc., you often create documents used by the firm's marketing department. Your challenge is to make sure marketing personnel understand how each New Age product should be sold and used properly, because many of the materials and processes involved in construction can be not only complex but also potentially hazardous to workers and customers. New Age has recently started shipping large amounts of its latest innovation: cellulose-based, blown-in insulation. You are new to this company and know only a few things about this new product—some good, some not so good.

Meant to replace foam insulation in homes and businesses, you know that cellulose-based insulation has many advantages over other types of insulation. First, it can be added to older buildings that do not have sufficient insulation or to new structures after the drywall has been installed. Second, the insulation is inexpensive; the material comes from old newspapers, with the only additive being a fire retardant and an insecticide to keep pests from nesting within a building's walls. Perhaps the greatest advantages of this type of blown-in insulation are that it does not give off fumes (unlike older foam types) and it takes it two to three times longer to settle than does foam insulation.

Some of the "not so good" effects of the insulation involve how it's installed. That's why you are meeting with Mark Godec, one of the insulation's inventors. You've been assigned the task of writing all the documentation to be included in the insulation's packaging, and he has expressed some concern about how you construct the information on the process of getting the material into the building walls.

"So the contractor has to drill holes in the wall? Big ones?" you ask Mark after you sit down in one of New Age's meeting rooms.

"At least four inches in diameter," he answers. "The hose they use to inject the insulation is about three and a half inches across. They need a hole at the bottom and the top of each run so they don't crack or blow out the inside wall."

You are taking notes on a legal pad, so you don't even look up when you ask your next question. "Do they ever drill on the inside of the house?"

"Not usually," he answers. "It's a lot harder to patch an interior wall than to temporarily remove a piece of siding or to do it before siding. Some contractors don't remove the siding. Instead, they put plugs over the holes, but we don't recommend that. It looks terrible."

"So how many holes can a homeowner expect to see?" you ask.

You know Mark likes to draw things out, so you aren't surprised when he grabs a piece of paper from your legal pad and begins sketching things out for you. He pushes his drawing (Figure 6.1) back at you, and you look at it while he continues.

"See how they need to drill the top and bottom of each empty space?" he says. "Some homes don't have fire stops, so there are fewer holes. Others have several stops, which means the contractors have to first find out where the stops are by using stud sensors."

"I get it," you say. "Okay, let's look at another aspect. How does the insulation get from its packaging, through the hose, and into the walls?"

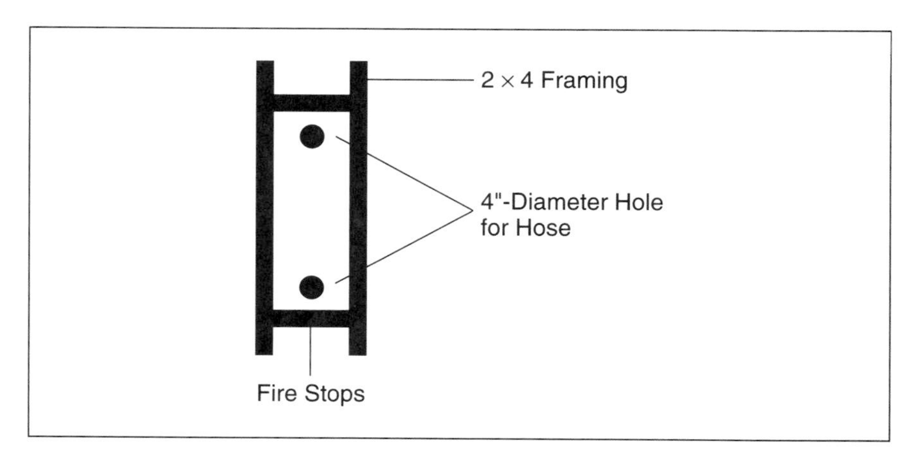

FIGURE 6.1 Sketch of Insulation Hose Holes

"Easy," Mark replies. "The insulation comes in plastic-wrapped bales. The worker opens up the blower machine hopper and puts the bale in, one at a time. They can't overload the hopper because that will jam up the blower blades."

"The blades?" you ask. "Where are they on the machine?"

Mark starts another sketch, this one of the blower machine (Figure 6.2).

"The machine has two sides: in front of the fan and behind the fan. The hopper is located behind the fan, so the insulation is sucked toward the fan blades. Before they get to the fan, they go through a series of rotating blades. The blades break up the insulation so it doesn't go into the walls in chunks."

"And so it can get past the fan blades," you note.

"Very true," he says, obviously glad you catch on quickly. "After that, the material is on the front side of the fan, so it's blown up the hose and into the wall. The worker knows the wall is full when the insulation will no longer travel up the hose. The fan will work harder, which trips a switch in the machine and stops the fan, so it doesn't overheat."

"Great," you say. "Any hazards to the installation?"

"Sure," Mark answers. "The workers have to wear masks. There's a bunch of fine particulate material floating around during this process, especially when they open up the hopper to add new bales or take care of a jam."

"It jams?" you ask. "Where? How often?"

"Oh, it depends," Mark answers. "Most of the time, the jam happens right in the rotating blades. The worker waits until the blades stop . . . oh yeah, when they open the hopper top it activates an automatic cutoff switch, which kills the fan and the blade motors. The operator has to take out all the insulation in the hopper, then use a metal rod, which comes with the machine, to poke the insulation through the blades. Sometimes, the jam happens at the fan—this means the operator has to open up the machine case and remove the insulation. The most common jam is in the hose, and

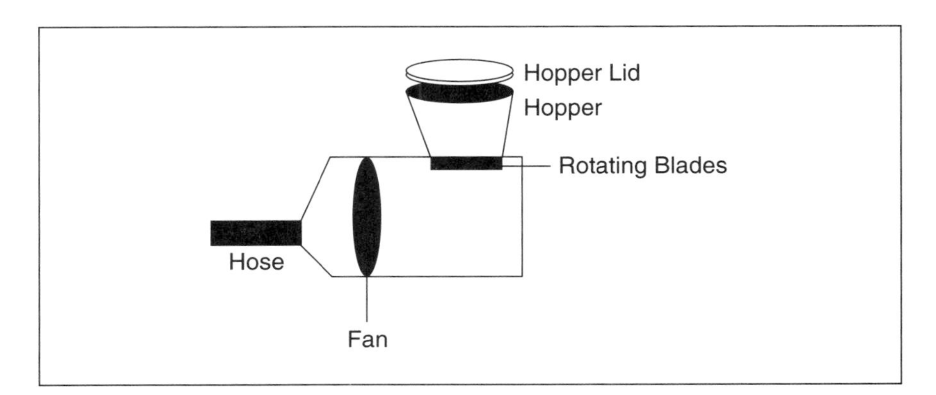

FIGURE 6.2 Sketch of Insulation Blower Machine

that's easy to take care of. You just unhook the hose, and pull out as much as you can. Then you run a long rod down the hose to push out what's in there. Most of the time, the jam's right at the coupling between the machine and the hose."

"Anything else I should know about the process?" you ask.

"Not really," Mark answers. "Oh, I almost forgot. I talked to Bill Smith yesterday, and he told me to tell you to make sure of a few things."

You can tell from his expression that Mark isn't very thrilled to be giving you this message from Mr. Smith, the vice president of New Age. "What things?" you ask.

"Well, it's just that he's sore enough about having the contractors wear masks during installation," Mark says. "He doesn't think customers will be too thrilled with the idea of masked workers swarming over their buildings."

"But the product is more environmentally friendly than any other," you say. "And our research has shown that the fine particles only happen during the blow-in."

"I know, I know," Mark replies. "And he wants you to make sure that you word the procedure on clearing jams very carefully. I guess he read about the lawsuits on snow blowers—you know, people clearing jams and losing their fingers in the blades. He doesn't want any ambiguities in the wording." "

"Understood," you say. "When do you want to see a draft of the install process?"

"As soon as possible," Mark says. "We need to get the new packaging materials out right away. We're not very happy with the current ones."

"Can I see what we have right now?" you ask.

"I'd rather you didn't," Mark says. "Start from scratch. We're ditching just about everything from the old materials anyway, and I don't want you to use anything from them just because they already exist."

"Sure," you say, not really understanding what the fuss is over. You and Mark leave the meeting room together and you return to your office to begin writing.

Just as you sit down, you remember an important fact that had not occurred to you before. When you had just started at New Age over six months ago, you had heard through the grapevine that another technical writer, working on the cellulose insulation almost a year before, halted the new product because of lung irritation. It turned out that the combination of the insecticide and fire retardant, when breathed in during the installation process, caused coughing and sometimes bronchial bleeding. New Age had almost folded but had found a better chemical combination that took care of the problem and managed to deliver the product before completely going under. The issue of lung irritation, however, was not entirely resolved, and the company was forced to include warnings that directed operators to wear standard filtration masks. No other problems with the insulation have been reported to your knowledge.

No wonder Mark was hesitant about these instructions, you think. *There's probably a lot of attention being paid to this product by the government regulators and potential buyers. I'd better make everything as clear as possible. But what about the mask warning and Smith's feeling it will hurt sales?* You begin jotting down notes, thinking that this project isn't as simple as it first seemed.

SUGGESTED NEXT STEPS

- Review the guidelines for creating technical processes presented in this chapter.
- Sketch out the procedure for blowing the insulation into the walls. Decide where in the process to start. What materials and equipment are required to begin? What needs to happen before the insulation is blown in? Justify your choices in your Solution Defense.
- Consider the ethical issues involved in how you word the warnings involved in this process. Discuss your decisions in your Solution Defense.

Where's the On-Ramp to the Information Superhighway?

CONSIDERATIONS

As someone who is learning to use new technologies as part of your college studies, you are often the perfect candidate for teaching your own new skills to others. In this situation, you are also saving yourself and others time by not repeating information. As you read the scenario, keep the following questions in mind:

- Think back to the first time you used the Internet, perhaps at your school's computer lab. What were your questions, frustrations, or mistakes during that experience?
- What is the skill level of your readers?
- What considerations, other than the Internet log-in procedures, are involved?

You are just starting your sophomore year at Technical University, a small four-year college in Hampton, Idaho. During your first year, you worked as a dishwasher in the University Center's kitchen. While the money kept you afloat during the school year, it wasn't the best job you'd ever held. This year, you have landed a position as a tutor in the campus computer lab.

As a tutor, you work in the lab, roaming between stations and helping students with their computing problems. Most of the time, you answer the same questions over and over, and sometimes you have to tell the same students the same thing several times. While you find this frustrating, you have not yet lost your patience, but you wonder how long you can remain completely polite.

The most common question you are asked concerns the procedure for logging on to the Internet and checking e-mail. The computer labs recently changed how students access their e-mail accounts because of security problems. You decide to talk to the computer lab supervisor, Melissa Swadley, about creating some kind of handout to give students seeking to surf the Net.

"That's a good idea," Melissa says. "We've been meaning to get something in writing, but no one has the time. If you want to tackle this, go ahead."

"Great," you say. "How much should I explain? Just the Internet or e-mail or both?"

"Both, definitely," Melissa says. "Maybe you can fit the instructions on the same page. You know, Internet on one side and e-mail on the other."

"Just one page?" you ask, thinking that the procedure is a bit complex, especially if you are going to include troubleshooting information.

"Yeah, keep it short and simple," Melissa says. "Besides, we don't have a big budget and I want to be able to photocopy this off whenever we need it. I don't have the money to take this to the campus print shop."

"I'll give it a try," you say. You sit down at a station and begin to work through the e-mail process, taking notes as you go. *This isn't all that simple,* you think, remembering how computer illiterate some of the students you have helped were.

Suggested Next Steps

- Review the guidelines for creating technical processes as outlined in this chapter.
- Go to your campus computer lab and learn how to access the Internet and your e-mail.
- With a lab attendant, tutor, fellow students, or the lab supervisor, discuss any problems that exist with Internet access and e-mail. Decide how you will handle these problems in your process, and justify your decisions in your Solution Defense.
- What kind of visuals will you use in your process? Research ways to take "screen shots" (pictures of the computer screen). Discuss your choices of visuals in your Solution Defense.
- How will you handle the length restrictions placed on this document? Justify your decisions in your Solution Defense.

Save the Planet

CONSIDERATIONS

In this situation, you have volunteered your writing skills for a unique project—making sure your company's new recycling project is successful. As you read this scenario, keep the following questions in mind:

- What are the various difficulties involved in the company's recycling program?
- What aspects of the project might motivate employees to actively participate?
- What process must you go through to have your recycling instructions disseminated?

You work as a public relations officer for Johnson Pharmaceuticals, a drug research and production company located in Freemont, Ohio. While you have only worked at Johnson for a few months, you already feel you've contributed to publicizing the company's advances in drug research and environmental responsibility. Last week, your supervisor, Mike Gert, sent you a very nice e-mail complimenting the press release you wrote about the company's new contract with a local paper recycling firm.

As you walk into the public relations office, you see a pile of cardboard boxes in the lobby. Pausing to look at them, you notice they have different labels: SHREDDER, PLASTIC, NEWSPRINT, GLOSSY, REGULAR. You see Mike Gert talking with a young man in cutoff jeans and a flannel shirt. Mike sees you and motions you to him. He looks confused.

"Maybe you can make sense of this," Mike says. "This young man is from EcoRound, the recycling company."

"I'm J. R. Wilson," the young man says, extending his hand to you.

"Nice to meet you, J. R.," you say shaking his hand. "What's the problem?"

"Well," says J. R. "You need to know how to deliver this stuff to us the right way. You see, our recycling machine can't sort everything out itself—it needs some human help. That's why we have these different boxes going to all of the offices. You fill them up with the right kind of paper material, and we won't have to charge the company extra for sorting at our plant."

"Okay," you say. "What kind of sorting do we need to do?"

J. R. begins to pick up boxes with different labels, explaining what goes in each. "Everything you want your company to shred before it ends up at our plant needs to go in the 'SHREDDER' boxes. Any paper with plastic attached, like window envelopes, needs to go in the 'PLASTIC' boxes. And then you have your specialty papers, like newsprint and glossy paper from magazines that need to go in their boxes. Everything else goes into the 'REGULAR' box."

"Okay," you say. "Seems pretty simple. Why have you brought us so many boxes?"

"We were told you needed them for each workstation," J. R. says.

"All those boxes won't fit at each workstation," Mike says.

"Why don't we have boxes for shredding and regular next to each wastebasket," you suggest. "Some of the areas share a wastepaper basket between three or four people. Then we could have the plastic, newsprint, and glossy boxes in the break area. I doubt most people will have much of that stuff on a regular basis."

"Good idea," Mike says. J. R. smiles and begins taking the boxes into the inner offices, placing them next to each wastebasket.

"Anything else?" you ask Mike.

"Yes, as a matter of fact," Mike answers. "Because you are now the one person totally up to speed on our new recycling efforts, why don't you write a memorandum to everyone, explaining how to sort everything properly?"

"Sure," you say, thinking that it will only take a few minutes. You and Mike are walking down the hallway toward your offices.

"You know," Mike says, "you may want to be very, very direct on your instructions. I overheard at a manager's meeting that any office that doesn't properly sort its recyclables will have to pay the extra sorting charge out of the office's regular operating budget."

"How much is the charge?" you ask.

"Well, if it's a great deal of material, we're talking a few hundred dollars. That might not seem like much, but our operating budget is so tight, I've been forced to end some of the perks, like making personal photocopies. Besides, if we cost ourselves money because we can't sort paper properly, that doesn't make us look very good."

You agree with Mike, and you also begin to see the potential for mistakes—like someone putting plastic bottles in the "PLASTIC" paper recycling box. "I'll get a draft of the recycling process memo on your desk as soon as possible," you say.

"Good," Mike answers as he steps into your office.

Suggested Next Steps

- Review the guidelines for creating technical instructions as outlined in this chapter.
- Carefully consider what tone to take when drafting this memo. How do you impress upon the employees the importance of properly sorting their paper? Justify your decisions in your Solution Defense.

- Incorporate some kind of visual in your document, perhaps one that gives examples of the types of material that belong in each recycling bin. Discuss your choices of this visual in your Solution Defense.

Soil Incineration: Ash Transportation

CONSIDERATIONS

For this portion of the soil incineration project, you must research and write about a specific part of the plant's operation—the removal and transportation of the toxic ash collected by the incinerator's pollution controls. After reviewing the information from previous soil incineration scenarios, read the following situation and keep these questions in mind:

- What are the political issues involved in transporting toxic waste?
- What pieces of information do you have concerning the incinerator's collection and shipment of ash, and what information do you need?
- Who are the readers for this document, and what is their familiarity with the project?

Your work on the soil incineration project is, thankfully, nearing completion. You have recently put the finishing touches on the technical description of soil incinerators, and you are feeling confident that the county board will receive enough information to be able to approve the project.

The public outcry concerning the soil incineration project has begun to die down; however, the organization created to oppose the project (Protect Our Land [POL]) has not ceased to make their concerns known to every government body involved in approving the necessary permits for the incinerator's construction and operation.

Today, you are putting the finishing touches on a report for another project when Jack MacGregor, your supervisor, enters your office. Jack looks grim; you anticipate yet another problem with the soil incineration project. "Got a minute?" he asks, sitting down.

"Sure," you reply.

"How far are you on the soil incineration report?" Jack asks.

"Not far," you reply. "The technical description is finished, and I was just starting to compile the information you requested for the township report. Why?"

"Well, it has to do with that township report," Jack says. "You need to do more than just overview the incineration process. It seems that POL has created some concern in the township trustees." Jack pushes a letter across your desk to you. It reads:

CEDAR TOWNSHIP BOARD OF TRUSTEES

Dear Donnelley Engineering Company:

As you are no doubt aware, the Cedar Township Board must grant several variances and approvals, in addition to those you must obtain at the county level, for the soil incineration project to proceed. We understand that your firm is handling all information requests concerning these approval processes.

It has come to our attention that the soil incinerator will produce toxic ash that must be transported out of the area to a toxic waste site downstate. We want to know how the incinerator company plans to transport this waste, and what the possible dangers are to our township's environment should an accident occur during transport.

Please answer the above questions, in writing, as soon as possible. The board wishes to review the information you provide before our next meeting.

Sincerely,

John Swanson

John Swanson
Chair, Cedar Township Board of Trustees

"Hhmm," you say after reading the letter. "I think the only thing we've said about the incinerator's by-products is that the material from the baghouse is stored in airtight drums, which are then transported to a toxic waste facility."

"What are the government regulations for transporting incinerator ash?" Jack asks. "I have a feeling that whatever we say we're going to do with the stuff, POL will be ready to compare it to what they think are legal transportation procedures."

"I'm not sure," you say. "I'll have to look into it. Why don't I gather together all the information on the processes and give it to you?" you suggest, feeling that the conversation isn't really bringing your company any nearer to handling the township's concerns. "I'll have to do some research on many different levels: any local ordinances concerning transportation of toxic materials, and of course the state regulations and the Environmental Protection Agency rules. This could take some time."

"Get right on it," Jack says. "I want a defined process for preparing and transporting the ash written as soon as possible. We don't want the township passing any new ordinances on us before we even get the chance to tell them what we're going to do."

"I understand," you say.

"Oh, and remember," Jack says as he is leaving your office. "All regulations are subject to interpretation. Keep your eye out for things we might not absolutely *have* to do. Maybe we can save our client a little money."

"I'll try," you say, not making any promises. *Isn't POL looking for just this kind of corner-cutting?* you think.

Suggested Next Steps

- Review the guidelines for creating technical processes as outlined in this chapter.
- Research the local, state, and federal laws and regulations on transportation of toxic materials. Assume that most of the ash will be trucked out of the incinerator site. In your solution defense, indicate where you found information on such laws and regulations.
- Consider the ethical dilemmas present in this situation. It is your job to help the client operate in the most profitable manner possible; however, you are dealing with laws and regulations that may conflict, contain vague language, and otherwise leave room for interpretation. In your Solution Defense, discuss how you have handled the issue of legal interpretation.
- Did you find potential "loopholes" in the law? Did the regulations sometimes recommend, but not require, certain things? Did you incorporate these recommendations into your process description as requirements, merely recommendations, or did you leave them out entirely? Justify your decisions in your Solution Defense.

CHAPTER 7

Resumes and Cover Letters

Resumes and cover letters are some of the most important pieces of communication you will ever write, and you may be surprised to learn how businesses treat these documents. When businesses advertise for new employees, they will more than likely receive a great number of applications from people who are not qualified for the position. Because your documents will probably be among hundreds, how do you make sure your resume and cover letter get the attention they deserve?

SEEING BOTH SIDES: JOB SEARCHING AND HIRING

When you are searching for a job, your first contact with the hiring company is only one part of the process—the job advertisement or "word" of an opening passed along to you by a friend, family member, or colleague. In fact, hiring an employee is, for a business, an involved, detailed process.

How Businesses Find New Employees

After identifying a need for a new employee, a business usually creates documentation that defines the employee's responsibilities, required qualifications, position in the company's hierarchy, and compensation. This documentation is called a *job description*. The advertisement for the position is based on the job description. After placing the advertisement, the company must then deal with the application process.

The typical process is for personnel employees to glance first through the cover letters and resumes and divide them into two piles—qualified and unqualified. They then reread the resumes in the "qualified" pile and start to rank the applicants according to experience, availability, and other factors particular to the open position. The top ranked applications for one position are then sent to the department or person responsible for hiring and working with the employee; the department then decides whom to interview. Every applicant receives a letter either requesting an interview or explaining why they were not appropriate for the position. During this entire process, different state and federal laws govern confidentiality and hiring procedures.

How Your Application Is Treated

This process does not leave enough time for anyone's resume and cover letter to be read closely until it reaches the interview consideration stage. While it may not seem "fair," consider the number of applications a business receives and the amount of time they have to devote to hiring the "right" new employee. You must construct your resume and cover letter in such a way that your qualifications are readily apparent to anyone quickly glancing over these documents.

IMPORTANT CONSIDERATIONS

The first chance you have for catching your reader's attention is when they go over your cover letter. When drafting your resume and cover letter, keep these important points in mind.

- *Your application documents will have several readers.* Your resume and cover letter will pass through many hands—personnel employees who simply sort the documents, those who decide who to interview, the interviewers themselves, and the company's higher-level management (who may have to approve the department's choice).
- *Your application will NOT be read thoroughly.* Typically, your resume and cover letter will receive a cursory glance to determine whether or not your basic qualifications match the job description. Count on having thirty to sixty seconds of reading time devoted to your documents until they receive more attention later in the hiring process.
- *Your readers need to be* told *why you are the "right" person for the job.* Resumes and cover letters are, essentially, arguments for why the company should hire you for the job over all other applicants. Don't be afraid to be direct. For example, you are competing for a job that requires extensive report writing. Instead of simply listing "extensive report writing experience" on your resume, be sure to add a state-

ment to your cover letter like "While working at Fairchild, I created several reports detailing the sales trends and marketing strategies of each product." Having detailed your experiences, your readers will have a better feel for your competency.

COVER LETTERS

Your cover letter is the document that details how your education and experience match the company's job description. When writing a cover letter, remember that you have several readers who will be spending very little time looking at your documents.

Cover Letter Format

Cover letters follow the same formatting rules as regular letters. You may choose either block or modified block format (see Chapter 2). Consider using expensive, thicker paper for your cover letter—its different feel may cause your reader to spend a little more time on your document. Avoid using dark-colored or patterned paper, as it will not photocopy well.

Examine the cover letter in Figure 7.1 on page 172 and carefully observe its construction.

Daniel Smith
100 Park Way
Grand Fork, Michigan 55555
616-555-5555

January 1, 1999

Human Resources Department
Technological, Inc.
100 Future Tech Drive
Grand Fork, Michigan 55555

Dear Human Resources Department:

Enclosed is my resume detailing my qualifications for the Software Programs Trainer position advertised in the December 26th *Daily Journal.*

Professional Background

I am currently employed as a trainer for a small computer company, teaching clients how to use basic software programs such as Microsoft Word, Excel, Access, PowerPoint, Publisher, and many other software products from Corel, Adobe, and Intuit corporations. Prior to becoming a trainer, I taught regular computer courses for Grand Fork State College for two years, thus gaining experience in teaching people with a variety of computer and technology backgrounds and aptitudes.

Special Qualifications

As this position requires extensive travel, let me assure you that I have no objection to being out of town for whatever length of time required. Also, I understand that you have a major client in Quebec, Canada—I am fluent in French, which may be of assistance to your company.

Please contact me after 6 p.m. at the above number to schedule an interview. I am also available at any time through my pager (555–1771). Thank you for your consideration.

Sincerely,

Daniel Smith

Daniel Smith

Enclosures

FIGURE 7.1 Cover Letter

Note these particular format considerations:

- *Return Address.* Be sure to include your name, address, and the phone number where you want to be reached. Sometimes you do not want prospective employers calling you at your current job. Also, note that the author in the sample letter added shading to his name. With today's word processing programs, it's easy to add lines or shading, design elements that can help separate your resume from others.
- *General Salutation.* If you know the name of the person or department handling the applications, use it in the salutation. However, companies often choose not to list an individual or department in their advertisement. Use "Personnel Department" or "Human Resources Department" in these situations. Do not use "To Whom It May Concern" or "Dear Sir or Madam" in a salutation.
- *Headings.* Because your readers are skimming your letter, use headings to indicate the kind of information contained in each paragraph. Headings serve the same purpose as the raised tabs on file folders—they allow your reader to quickly locate the information in which they are interested.
- *Simple Complimentary Close.* "Sincerely" is the best way to close a cover letter. It's simple, direct, and the most common form for a business letter.
- *Enclosures Indication.* Because you are including your resume, references, and any other pertinent application material, indicate these additions by typing "Enclosures" on the last line of your letter.

Cover Letter Content

In the cover letter example provided in Figure 7.1, note the following content elements:

- *Immediate Orientation.* The first paragraph immediately alerts the reader to the purpose of this letter. Because some companies may have advertised several open positions, be sure to indicate when and where you saw the advertisement and for which position you are applying. Avoid unnecessary comments in this introductory paragraph, such as "I noticed your advertisement in the paper" or "I heard you had a position opening soon."

- *Overview of Most Important Information.* Again, remember that your reader is skimming this letter. Be sure to write about those parts of your education and experience that directly relate to the position for which you are applying.
- *Mention of Special Qualifications.* Make your resume stand out from the rest by including information that may make you more desirable to the company, such as your willingness to travel or relocate. In the sample cover letter (Figure 7.1), the author showed that he did his homework—he knew the company had a major client in a French-speaking region of Canada and pointed out how his language skills match the company's needs.
- *Method of Contact.* Most businesses would rather call or write you if they want an interview. Sometimes the hiring process takes longer than expected—imagine a business' frustration if all the applicants called to request interviews! Give your prospective employer a number and time to reach you and make a commitment to being available every day.

RESUMES

Your resume lists all your experiences and qualifications. Remember that you do not have to list *every* employment or educational item. You want to call attention to those that particularly match your prospective employer's needs. When creating your resume, pay attention to format details. Your reader should be able to notice the important information quickly and easily.

Resume Format

There are several ways to format a resume. Decide what information should receive the most attention. Were your job titles or places of employment more impressive? Is your education your strongest point for this particular job? Figures 7.2 on pages 175–176 and 7.3 on pages 177–178 demonstrate two different resume formats: the first emphasizing education and job titles and the second job experience and place of employment.

Daniel Smith
100 Park Way
Grand Fork, Michigan 55555
616-555-5555

OBJECTIVE

To obtain full-time employment utilizing my teaching experience and knowledge of current computing technologies.

EDUCATION

Bachelor of Science, Computer Information Systems

Northern Michigan University, 1994

3.8 Cumulative Grade Point Average

Teaching Minor

PROFESSIONAL DEVELOPMENT

Computer Technologies Training Certification Programs

Midwest Technology Training Academy, 1994

Certified to teach all Microsoft, Corel, and Adobe software products

David Karnel Seminars

Completed the following seminars: *Training the Trainer, Working with Difficult People, Overcoming the Fear of Technology, Using Multimedia to Enhance Computer-Oriented Training.*

EXPERIENCE

Trainer

Future Tech, Incorporated 1997 to present

Grand Forks, Michigan

Trained clients in all Microsoft, Corel, Adobe, and Intuit software programs. Developed curriculums and assisted in the training of new trainers.

Instructor

Grand Forks State College 1995–1997

Grand Forks, Michigan

FIGURE 7.2 Resume Emphasizing Education and Job Titles

Evening and weekend course instructor for basic computing classes on various word processing, spreadsheet, and database software.

SKILLS

Fluent in French-Canadian language and culture as I was born in Montreal, Quebec—I still retain dual U.S.-Canadian citizenship.

Proficient in all current versions of Microsoft, Adobe, Corel, and Intuit major software programs. Experienced web-page designer with HTML and Java programming skills.

REFERENCES

Mike Schwinger
Director of Training
Future Tech, Inc.
616-555-5333

Dr. Carey Greyhawk
Department Head
Grand Forks State College
616-555-5455

FIGURE 7.2 Resume Emphasizing Education and Job Titles *continued*

Daniel Smith
100 Park Way
Grand Fork, Michigan 55555
616-555-5555

OBJECTIVE	To obtain full-time employment utilizing my teaching experience and knowledge of current computing technologies.
EXPERIENCE	**Future Tech, Incorporated** Grand Forks, Michigan 1997 to present *Trainer.* Trained clients in all Microsoft, Corel, Adobe, and Intuit software programs. Developed curriculums and assisted in the training of new trainers.
	Grand Forks State College Grand Forks, Michigan 1995–1997 *Instructor.* Evening and weekend course instructor for basic computing classes on various word processing, spreadsheet, and database software.
SKILLS	Fluent in French-Canadian language and culture as I was born in Montreal, Quebec—I still retain dual U.S.-Canadian citizenship. Proficient in all current versions of Microsoft, Adobe, Corel, and Intuit major software programs. Experienced web-page designer with HTML and Java programming skills.
EDUCATION	**Bachelor of Science, Computer Information Systems** Northern Michigan University, 1994 3.8 Cumulative Grade Point Average Teaching Minor
PROFESSIONAL DEVELOPMENT	**Computer Technologies Training Certification Programs** Midwest Technology Training Academy, 1994 Certified to teach all Microsoft, Corel, and Adobe software.

FIGURE 7.3 Resume Emphasizing Job Experience and Place of Employment

David Karnel Seminars

Completed the following seminars: *Training the Trainer, Working with Difficult People, Overcoming the Fear of Technology, Using Multimedia to Enhance Computer-Oriented Training.*

REFERENCES

Mike Schwinger
Director of Training
Future Tech, Inc.
616-555-5333

Carey Greyhawk
Department Head
Grand Forks State College
616-555-5455

FIGURE 7.3 Resume Emphasizing Job Experience and Place of Employment *continued*

In these examples Figures 7.2 and 7.3, note the following format elements:

- *Name, Address, and Phone Number.* Be sure to include this information at the top of your resume. With modern word processors, you may apply design features, such as lines or shading, to make your resume more attractive.
- *Floating Headings.* Running your major headings by themselves along the left side of the page allows your reader to quickly skim your resume and read only that information he or she wants. It also allows readers to make notations in this ample white space.
- *Type Style.* Whatever your type style scheme, be consistent. In these examples, the major headings were capitalized and bolded, with subheading mixed capitalized and bold. Softer emphasis, in the form of italics, was applied to job titles in Figure 7.3.
- *Paragraph Format.* Both these examples used block paragraph format, which again lends itself to skim-reading.

Resume Content

While the way your resume looks is very important (because it serves as your prospective employer's first impression of you), what your resume says and how it says it are equally vital. These elements should be part of every resume:

- *Personal Information.* Provide your name, address, and the phone number where you wish to be reached. Beyond this information, you are not required to divulge anything else about yourself. In fact, laws prohibit you being required to provide your age, gender, marital status, or racial background in certain employment situations.
- *Career Objective.* Let your employer know what you're looking for. Remember to match your career objective with the realities of the position for which you are applying. For example, if the advertisement is for a permanent part-time job, you may be hurting your chances if your objective mentions permanent full-time employment. While that may be your goal, mentioning it now may not be effective.
- *Employment Experience.* List the jobs you have held, from most to least recent. If you have been employed in many different places and positions, provide only those that relate to the kind of work the new job would require. Include your job title, the employer's name and lo-

cation, and the period of time you worked there. Write a brief description of your responsibilities (again, remember to match them as closely as possibly, while remaining accurate, to the kind of responsibilities you would hold if hired).

- *Education.* For college students and graduates, your higher education information should always appear on your resume. Provide the name of the degree you received followed by your major, the name of the institution, and the year you graduated. (*Note:* If you have not graduated, write "To Be Conferred" followed by your expected graduation date.) If your grade point average is above a 3.0, include it here, and if your minor area of study relates to your prospective employment, list it as well.
- *Professional Development.* This heading is for another kind of education—that which takes place outside a formal institution. Any conferences or seminars you have attended or programs that resulted in certifications are listed here.
- *Skills.* Many employers like to see a list of the various skills you will bring to their company. Expertise in computer programs, other technologies, and foreign language fluency can help put you one step above an applicant with similar qualifications. Avoid common and trite phrases such as "excellent people skills" or "good interpersonal relationship skills." Include this kind of information in your cover letter if you feel it's important to mention.
- *Volunteer Activities.* While we did not include this heading in our examples, employers generally like to see proof of your commitment to the community in which you live and work, and in what direction you expend your off-work energy. Again, this kind of activity can be a deciding factor between you and a competitor with a similar professional background.
- *References.* Provide the names, titles, organization, and phone number for at least two people who can speak about your past performance and abilities. Avoid using the phrase "references provided upon request" because this requires your prospective employer to do the legwork. Remember to ask your references if you may use their names, and tell them what kind of companies will be calling and what kind of jobs you are seeking.

COMMON EMPLOYMENT CORRESPONDENCE MISTAKES

What employers expect from a resume and cover letter has not, essentially, changed in the past decade; however, job seekers have used various techniques to get their resumes noticed, with varying degrees of success. Here are some common mistakes, many of them outdated methods, that job seekers make.

- *Using Brightly Colored Paper.* While bright orange, yellow, or blue might stand out from the others in a resume pile, stick to professional colors, such as white, off-white, light ivory, and light grey. Your resume will probably be photocopied, so try this paper test. Print your resume on the paper you want to use, and then photocopy it. If the quality is not almost as good as the original, find better paper. (*Tip:* Stay away from dark-colored papers or those with dark-colored flecks in them. These photocopy poorly.)
- *Long Paragraphs.* You may have much to tell your employer in your cover letter. Your first consideration, however, is the amount of time someone will actually spend *reading* your letter. Short paragraphs and the use of headings allow for accurate skimming.
- *Typographical Errors.* Your resume claims you pay close attention to detail, but your indents are off and you have three typos. How will your prospective employer evaluate your "close attention to detail"? Resumes and cover letters must be 100 percent free of formatting, spelling, or any other type of error. Ask friends, family members, and colleagues to proofread your employment correspondence for you.
- *Improper Salutation.* Nothing might sink your application faster than using "Dear Sir" as your greeting when the person hiring you is female. You might be perceived as not willing to work with women when you really have no such problem. Many older textbooks claim this is an acceptable salutation if the person's name is not known. Use "Dear Human Resources" or "Dear Personnel Department" if you do not have any other information.
- *Lengthy Cover Letters and Resumes.* Keep your cover letter to one page. Your resume should not exceed two pages. If necessary, list your references on a separate page. When faced with having to shorten up your resume, take a look at the job advertisement and review any research on the company. Keep only those items that directly relate to the position for which you are applying.
- *Faulty Parallelism.* Stick to the same kind of verb tense and sentence construction, especially when describing your past employment. Here is an example of faulty and correct parallelism.

Faulty Parallelism	*Correct Parallelism*
My duties included staff training, product updates, and overseeing cash handling.	My duties included staff training, product updates, and cash handling oversight.

Exercises for Classroom Discussion

1. Divide a piece of paper in half. On the left side, list each job you have ever held. On the right, list all the things you learned from that job. You will quickly realize that you have qualifications you may not have considered. For example, if you have worked for the family dairy farm, you have learned many things—how to arrive at work on time (even in the early morning), how to handle crises (such as sick animals), how to operate specialized equipment, and how to follow detailed procedures. These are all very marketable job skills.

2. Collect three job advertisements from newspapers, magazines, the Internet, or your college's employment office. In a memo to your instructor, explain what you think the employers are looking for and how your current experiences and qualifications would make you a good candidate for each job.

3. You are the owner of Future Tech, a small computer company. Recently, your business began designing web pages for area companies, and you are now deluged with projects. You realize you cannot satisfy the demand on your own; however, you have very little extra money to hire another employee. You decide to hire someone from Technical University, a local college, as an intern. You know an intern might not have all the Internet, graphic design, programming, and writing abilities you need, but you are willing to train someone to do it your way. You do want someone who has at least tried to make a web page using an HTML editor of some type. Write the job advertisement for this internship position.

4. You are a sophomore English major at Technical University, and you see the job advertisement for a Web Design internship at Future Tech, a small, local computer company. While you know more about the Internet as a surfer than as a designer, you are interested. You've used Netscape Navigator's editor to create a simple web page, you managed to create links to other sites, and even made a small animation of the word *hello* using the Java programming language. You know how to scan images and use a photo editor. Your main concern, however, is your lack of experience in the computer field; your past jobs were writing for the college newspaper (*The Tech Weekly*) for the last year and working the graveyard shift waitressing at Big Al's Diner while you were in high school. Deciding that, because the position is an internship, the employer isn't looking for enormous experience, you write a resume and cover letter for the position.

5. Find an advertisement for a position you would like to have while working your way through college. Write a cover letter and resume for that position. Give the advertisement, your cover letter, and your resume to a classmate. The classmate will write down a list of questions the employer might ask the job candidate, and will try to find answers to those questions in your cover letter and resume. When your classmate is finished, discuss how many questions could be answered by your documents, and how well those questions were addressed.

SCENARIOS FOR RESUMES AND COVER LETTERS

The Boss's Son

CONSIDERATIONS

Personnel departments handle various employee issues, including hiring, firing, employee complaints (such as discrimination and sexual harassment) health insurance, and retirements. In this situation, you are put in a difficult position—a relative of your employer has applied for a position in the company. As you read this scenario, keep the following questions in mind:

- What are the personal, social, and political relationships between each of the people in this situation?
- Is there anything about John Martin, Jr.'s application that leads you to believe he expects the position?
- What are the implications of sending the young Mr. Martin's application further along in the process?
- What are the possible outcomes if you do not recommend the application for further consideration?

As the assistant manager of the personnel department at Rayco, a toy manufacturing company, you deal with many difficult situations. In fact, you've just left a meeting about whether or not to fire an employee over excessive absences, and you feel more than a little drained. The employee in trouble is a single mother with two children and claimed to be staying at home to take care of her youngest, who was very ill.

You are still thinking about the meeting, a little pleased that the employee was only put on probation, when Kelly, your boss, knocks on your office door.

"Tough morning, wasn't it?" she says as she sits in the chair across from you.

"Yes," you reply. "I'm still not sure we helped that woman much. She's bound to miss more work because of her children."

"Well, we did the best we could, considering," Kelly answers. You know she is referring to John Martin, Rayco's vice president. Mr. Martin has been pressuring the personnel department to review all employees and recommend firing or laying off those who are the least productive. It seems that Rayco isn't making as much profit as it used to, but you know the company is not in any serious trouble.

"Well, I didn't come here to talk about this morning," Kelly continues. "We have a slightly bigger problem on our hands."

"What is it?" you ask.

"I just went through all the applications for the marketing assistant position, when I came across this," she says as she pushes a folder across your desk.

You pick it up and see it is a regular cover letter, which reads:

John Martin, Jr.
10 Dinero Lane
Big Rapids, MO 43332

Rayco, Inc.
1000 Rayco Drive
Big Rapids, MO 43331

Dear Sir:

Allow me to introduce myself. I am John Martin, Jr., and I am interested in the Marketing Assistant position that my father, the Vice President of Rayco, recommended to me.

Please review my considerable qualifications and call me at your convenience to schedule an interview. I may be reached at 555-5143 after 6 p.m.

Sincerely,

John Martin, Jr.

John Martin, Jr.

"Where's his resume?" you ask.

"Oh, I didn't put it in there. Mostly because there's nothing to it. This guy is fresh out of college—four year degree in Public Relations with a 2.2 grade point average. No job experience, because he didn't work at all during the last five years when he was in school."

"Did he do any volunteer work, or was he part of a student organization?" you ask.

"No. Nothing," Kelly answers.

"Who did he put down for his references?" you ask.

"Just one name," Kelly answers.

"Let me guess—his father's," you say.

Kelly just smiles sarcastically, and you know you're right.

"Do you think Mr. Martin knows anything about this?" You hope that John Jr. sent the application without telling his father what he'd done.

"Oh yes," Kelly said. "Right after this morning's meeting, Mr. Martin took me aside and asked how our search process was going for the marketing position. He said, and these are his words, that he had 'sent a very qualified candidate our way' and was sure this candidate would 'be treated very well'."

"This is just wonderful," you groan. "Martin's been all over us about getting rid of unproductive people, and now he wants us to hire his unqualified son!"

Kelly was silent for a minute, obviously thinking. "You know, I never thought about this that way. . . ," she mutters.

"What way?"

"Martin could retaliate if we don't hire Johnny Jr.," she says. "He could say we haven't done enough to improve worker productivity."

"He wouldn't dare," you say. "We've been very careful to document all our evaluations, and we've already let more than a dozen line workers go. I think if he tried that tactic, it would be very apparent from the records that he's wrong."

"I hope so," Kelly said. "Now to the larger question—how are you going to handle this?"

" Me?" you say.

"You were supposed to handle the initial cut and report for this position," Kelly reminds you. You realize that the part of your job you find so easy—simply sorting out the applicants who are wildly unqualified and sending the rest to the hiring department, along with the initial recommendations for top candidates—had now become very complicated.

"So I don't remove him from the applicant pool," you say. "That much seems obvious."

"But do you recommend him as a top candidate?" Kelly asks.

"You tell me," you say.

"Oh, no, I'm not going to tell you how to do your job," Kelly says, laughing. "But," she added, becoming very serious, "I will support whatever decision you make."

"Thank you, I think," you say.

"Good luck," she answers as she leaves.

Later that afternoon, you receive the entire application file for the Marketing position. After reviewing all the applications, you know that there are only five people with the right qualifications for the position. *Good,* you think. *That makes for the basic "here are the only qualified applicants" report.* You come across John Martin Jr.'s resume and agree with Kelly's assessment. Under normal circumstances, you would never consider including this kind of applicant as a serious candidate for a position, but now you wonder what you should do.

SUGGESTED NEXT STEPS

- Review the process for working through ethical dilemmas as outlined in Chapter 2.
- Review the criteria for good resumes and cover letters provided in this chapter to evaluate John Martin, Jr.'s application.
- Decide what course of action you should take.
- Write to the appropriate person, defending your decisions on this round of applications.

Internship Adventure

CONSIDERATIONS

In this scenario, you can really be yourself. As you read this situation, keep in mind the following questions:

- Why are internships of value to college students?
- What type of experience would help you obtain your career goals?
- What resources are available at your college or university to assist you in finding an internship?

It is your sophomore year of college, and you have no special plans for the summer. You've been toying with the idea of applying for a summer internship somewhere—one of your friends recently landed a summer position as a journalist assistant in Washington, D.C. A friend of a friend is traveling to England to work on an archeological dig, and an ex-roommate is going to New York to work for MTV. The prospect of spending another summer doing the "same old thing" doesn't appeal to you, especially given the exciting things your friends will be doing. You want some kind of adventure.

You decide to call up Melissa, the friend who will be working in Washington, D.C., and ask her how she got the job. After exchanging hellos, you get right to the point.

"Listen, Mel, I was wondering if you could help me out."

"With what?" Mel answered.

"Well, I was kind of thinking about getting an internship for the summer, and I wondered how you went about it," you say.

"Not much to it, except for some legwork and research," Melissa answers. "I went down to the university's employment office and started asking some questions. It seems they had information on internships, like what was out there and who to write."

"Great," you say. "That sounds easy."

"Wait a minute," Melissa says. "It wasn't a piece of cake, you know. I had to do some research on the places where I wanted an internship and then write them each letters. And I had to make up my resume and send it out to them. It took me a couple of weeks to get it all done."

"Oh," you say.

"So what kind of internship are you going for?" Melissa asks.

"I don't know just yet. I guess I'll have to see what's out there," you answer.

After hanging up the phone, you grab the campus phone book and look for the number for the university's job search office. You find it but hesitate. What do I really know how to do? you ask yourself. I haven't had that much experience!

Suggested Next Steps

- Make a quick list of all the jobs for which you might be qualified when you graduate, and then another list of all the things you've done that an employer might want to know.
- Call your school's job search office and ask about internships to get the process rolling.
- Review the guidelines in this chapter for writing your resume and cover letter. Create your resume and cover letter for the internship.

Intern Search

Considerations

In this scenario, you find yourself in need of some good help—otherwise, you will have a tough time doing well at your job. Your company also does not want to pay much, if anything, for another employee, so you decide to look for a college student in need of experience. As you read the following situation, keep these questions in mind:

- What are the tasks for which you will need assistance?
- What process must you go through to hire an intern?

- What are the necessary things you must do to gain approval for your choice of intern?

As an assistant manager in BioTech, an environmental engineering company, your responsibilities are mainly in-depth research into whatever project your manager is working on. BioTech's projects range from testing wells for contamination and creating soil quality reports to establishing the environmental impact of any new construction projects. Most of what you do is reading and interpreting federal environmental regulations and determining if they apply to the current project.

In a staff meeting, you admit to feeling overwhelmed by your work. Congress is in session, and the changes to environmental regulations are happening faster than you have time to read and understand the new laws. After the meeting, Carey Blackeagle, your supervisor, asks to speak with you. You meet in her office. "Do you need help?" she asks.

"Well, yes," you say. "I'm not sure if any of the regulation changes will affect our work with the Column Oil construction project." Column Oil is an older refinery currently attempting to upgrade its facility, because the Environmental Protection Agency is threatening stiff fines. Your company is helping them find ways to clean up existing pollution before the groundbreaking for a new addition to the refinery begins.

"What about hiring a college student to help you out?" she asks. "We haven't had an intern around here for a while, and I'm sure you could find someone willing to work part-time for the next few weeks."

"Sure," you answer, grinning. "Interns are usually not hard to find."

"Be sure to tell me whom you hired and what qualifications he or she has," Carey tells you. "I am reserving final approval of your intern. I also want the intern to have a specific job description."

"No problem," you say. "Thanks."

"You're welcome," she says, smiling. Much more relaxed, you go back to your cubicle and look up the number of the student employment office at nearby Technical University. After several phone transfers, you speak to Constance Wallets, an off-campus employment coordinator. She tells you to send her a short internship description, including desired qualifications, hours, and pay; she will post it in the university's internship bulletins and recruit students herself.

You write to Ms. Wallets, with information you feel will attract qualified applicants. You don't want to sort through a stack of resumes from unqualified people.

Several days later, the personnel secretary gives you a stack of resumes and cover letters. You leaf through them and pull out three that look good. Only one seems like a very good candidate to you, and you decide to include Carey in the interview, so you can get her approval right away if the intern is a good match. You know that Carey likes to be well informed before any interview, so you fire up your computer to write her about this promising candidate.

Suggested Next Steps

- Review the following three resumes.
- Decide on one candidate that you feel would be best qualified and, therefore, the most important to interview.
- Write to Carey to invite her to the interview and include any information you feel is relevant and appropriate for this situation.

Candy Alcove
210 West College
College Town, NY 55515
(931) 555-1111 (W) (931) 555-1212 (H)

Objective

To obtain valuable work experience in the environmental field.

Education

Northern University
College Town, NY
Bachelors Degree in Liberal Arts, to be conferred in August, 1998.

Interests and activities

Student Environmental Association (SEA) Treasurer.

Work experience 1994—present

Northern University
College Town, NY
Research Assistant. Assisted the head of the Biology Department with various research projects.

Hobbies

Ultimate Frisbee, fishing.

References

Provided upon request.

Bradley Campo
112 Van Antwerp Hall
Newton, NY
(906) 555-1112 (W) (906) 555-1211 (H)

Objective
To obtain a rewarding internship experience employing my research skills.

Education
Technical University
Newton, NY
Bachelor of Science, Biology. 3.8 GPA. December 1997 graduation expected.

Work experience 1992–1995
Bear Butte National Park
Bear Butte, Montana
Youth guide. Coordinated nature experiences for visiting youth, led extended camping trips focusing on environmental educational experiences.

Volunteer experience
Sierra Club. Served as president for four years. Coordinated "Endangered Species—Human" annual educational activities at technical university.

Awards received
outstanding biology undergraduate award.
outstanding member (sierra club)
technical university best program award

Interests and activities
camping, computers (Internet), and mountain climbing

References
Bob Cooley, Biology Department Head, 222 Bald Hall, Technical University, 555-2222.

Carl Shaw
188 West Second Street
Almherst, NY 55988
931-555-1234

June 7, 1999

Personnel Department
BioTech Engineering, Inc.
1000 Technical Street
Brighton, NY 65966

Dear Personnel Department:

Constance Wallets mentioned your internship opportunity to me on my last visit to her, and I agree with her assessment that an internship experience will help me as I change careers.

As you can see from my resume, I have extensive wastewater management experience. While I am not very familiar with the environmental engineering field, I am eager to build upon my knowledge of chemical testing and analysis.

I am sure you will agree that I am a more seasoned worker than many of your applicants, and will have no problem "fitting in" a professional setting. I look forward to bringing my experience to BioTech's engineering team.

Please contact me for an interview at your earliest convenience. I am anxious to begin working for you.

Sincerely,

Carl Shaw

Carl Shaw

Carl Shaw
188 West Second Street
Almherst, NY 55988
931-555-1234

Objective: To obtain an Internship in the environmental field

Professional Experience

City of Almherst August 1986 to 1993

Wastewater Treatment Technician. Conducted pH, toxicity, and other quality tests on treated effluent waters for the city water treatment system. Assisted the City Manager and Wastewater Treatment Manager with budget and capital expenditure proposals and customer billing rate proposals. Created community awareness program to decrease phosphate emissions and general wastewater emissions.

Jackson State Prison May 1975 to August 1986
Jackson, Michigan

Wastewater Treatment Assistant Technician. Worked in conjunction with Treatment Technicians to maintain wastewater emission qualities for this closed-system wastewater treatment facility. Conducted pH, toxicity, and other quality tests.

Education

Technical University
Newton, NY
Major: Zoology. GPA: 3.0, sophomore standing. 1970 to 1975

Western Michigan University
Kalamazoo, Michigan
Associates Degree, Wastewater Management.

References

Robert Whissler
Manager, Marquette Wastewater Treatment Plant
931-555-4455

Dr. Herb Gronk
Jackson Maximum Correction Facility Wastewater Treatment
313-655-1234

Barbara Ellen

1006 West Magnetic • Trighton, NY 59288 • Telephone: 931.555.5555

June 7, 1999

Personnel Department
BioTech Engineering, Inc.
1000 Tech Street
Brighton, NY 65966

Dear Personnel Department:

I was excited to hear about your position for an intern. I have been a research assistant for SBC Environmental, and would like to experience another firm's methods for conducting environmental assessments.

I have lived in this community ever since moving here to attend Tech University, and am looking forward to securing a full-time position in the area upon graduation. I am sure that gaining the experience and skills BioTech has to offer will assist me with this goal.

Please feel free to contact me at the above number to schedule an interview. I am looking forward to further discussing how I may be of assistance to your company.

Sincerely,

Barbara Ellen

Barbara Ellen

Barbara Ellen

1006 West Magnetic • Trighton, NY 59288 • Telephone: 931.555.5555

Objective

To further my education in the environmental engineering field.

Education

TECHNICAL UNIVERSITY
Bachelor of Science in Biochemistry, to be conferred May 1996

Experience

SCB ENVIRONMENTAL, INC.
Research Assistant, 1992 to present
Responsible for maintaining records on current environmental legislation. Conducted secondary research for Manager of Environmental Engineering as projects demanded. Made on-site visits and assisted with report drafts.

OLIE LIBRARY, TECHNICAL UNIVERSITY
Library Assistant, 1991–1993
Returned books to stacks, assisted librarians with cataloguing duties, assisted students with their research needs.

Skills

- Capable Internet user with own e-mail ID.
- Familiar with Microsoft Word and Excel applications.
- Experienced researcher, knowledgeable about information sources in the area.

References

LAURA PRICE
Manager of Environmental Engineering
SCB Environmental, Inc.
906.555.2525

CARL ROBARS
Librarian
Technical University
906.555.2255

The Judiciary Opportunity

CONSIDERATIONS

In this situation, you are pursuing a position in an influential branch of student government. As you read this scenario, keep the following in mind:

- What process must you follow to obtain an interview for this position?
- What experiences and skills do you possess that will make your application competitive?
- What are the advantages to obtaining this position?

You are a sophomore at State University, a medium-size college in Arizona. So far, you've done fairly well, but you realize that your chances of securing a good job after college depend on more than just your coursework. Recently, you attended a campus-wide seminar on how to gain valuable experience through volunteerism and serving on campus committees. You haven't had a chance to look through the packet of information given out at the seminar, and because your favorite TV show isn't on for another few minutes, you decide to go over it.

You glance through all the pamphlets from various student volunteer organizations, but none catches your eye. Leafing through sheets of colored paper describing campus committee openings, you do notice one.

MAKE THE TOUGH DECISIONS

The Faculty-Staff Judiciary has one opening for a Student Member. The Judiciary is primarily responsible for deciding the fate of students who have been charged with violating the Student Code of Conduct. The Judiciary meets once every two weeks, unless there are no cases to be heard.

Applicants must be at least sophomore status and be free of any disciplinary probation, university or otherwise. Send a cover letter and resume to "Judiciary Post, Box 10a, Dean of Student's Office." All applications will be reviewed by current Judiciary members and a representative of the Dean of Students office.

You decide that employers would certainly appreciate someone with this kind of experience. Making tough choices with serious consequences seemed a regular part of the Judiciary's work. You reach for the phone and call up your friend John, who is a year ahead of you. "Hey John," you say.

"I'm thinking about applying for the Faculty-Staff Judiciary. What do you know about it, besides what they do?"

"Not much," John says. "Except that there's stiff competition for those committee seats. I've tried twice myself with no luck."

"Really?" you say. "Why's it so tough?"

"Everyone wants on that committee," John says. "It's really intense. You get to hear cases, like students who have been charged with plagiarism, or stealing, or having keg parties on campus. It's like being a judge, really. You ask the defendants and plaintiffs questions, and then the committee has a closed-door session to discuss the case. Your vote carries as much weight as anyone else's."

"How many people are on this committee?" you ask.

"Seven. Four faculty and three students," John answers. "I don't know who's on it right now, though."

"Do you know why you didn't make it?" you say.

"No," John says. "But I've been told they are very, very careful about student members. Confidentiality issues, you know."

"Sure," you say. After thanking John for the information, you hang up and start digging through your files for your old resume. You've already decided to apply, and you want to make a new resume. *I wonder what I could say in my cover letter to get them to trust me?,* you think as you finally locate the right file.

Suggested Next Steps

- Review the process for creating your own resume and cover letter as outlined in this chapter. Make sure you have a complete listing of all your skills and work experiences.
- Taking into consideration the needs and attitudes of your potential readers, write your cover letter and resume for the Student-Faculty Judiciary.
- Have at least two people read and comment on your resume and cover letter before writing your final draft.

Soil Incineration Case: Research Intern Wanted

Considerations

In this portion of your ongoing soil incineration project, you find yourself under tight deadlines, with only two hands and twenty-four hours a day to complete too many projects. Because this is a sensitive project, keep in mind the following questions as you read this scenario:

- What are the potential benefits and hazards of hiring an intern or part-time employee to assist you with your work?
- What qualifications and skills are you looking for in an intern?
- What are Jack's concerns about hiring someone to help you?
- What is the procedure you must follow to hire someone?

Your projects at Donnelly Engineering are piling up. In addition to the soil incinerator project, you have three other time-consuming projects to complete within three months. Because the soil incinerator documentation is your top priority, you're wondering how you will complete everything on time.

After the monthly staff meeting, you ask your supervisor, Jack McGregor, to stay behind so you can mention your concerns. "I have a pretty good handle on everything except for the extensive research needed for the soil incineration project," you say. "I was wondering if we could bring on an intern for the next few months. Maybe give some college student a chance for some extra money."

"I don't know," MacGregor said. "This is a pretty sensitive project, and I'm not sure how I feel about some part-time person being very involved."

"I'm sure we could find someone who'd be careful," you say. "And after all, it's getting harder and harder for me to get out of the office and into the library. A student would be perfect—familiar with the college's resources and on campus much of the time."

"True," Jack said, looking thoughtful. "You know, I could use some research help on a few of my projects, and I can't seem to get away from here, either."

You smile hopefully. "Why don't we share an intern?" you ask.

"Sounds good to me," Jack answers. "I've got some professor friends at the university who could recommend some good students."

"Great," you say. "Anything I can do to help?" You really want to be part of the hiring process, because you will be working with this person.

"Yeah," Jack says. "Why don't you interview the candidates?"

"Sure," you answer.

"I'll be sending some students your way by next week," Jack promises.

True to his word, you have received calls from two students requesting interviews for the internship. You have scheduled meetings with both this afternoon. Jack's contacts at the university produced just these two candidates; however, Jack assured you they come highly recommended.

Carrie Smith, the first candidate, arrived. After exchanging greetings, she sits on the other side of your desk. She hands you her resume, which reads:

Carrie Smith
1010 Fitch Way
Hedgebrow, MY 55555
906-555-3563

OBJECTIVE

To obtain practical research experience prior to graduation

EDUCATION

Bachelor of Arts Degree
English Major, Conservation Minor
State University
Expected Graduation: May, 2000

EXPERIENCE

American Finance & Mortgage
Hedgebrow, MY

Worked two years as a clerical assistant in this busy accounting and brokerage firm. Performed typing, filing, and reception duties.

ACTIVITIES

President, Student Research Association
Volunteer, Habitat for Humanity

REFERENCES

Prof. Jo Erickson
Dept. of English
555-3613

Ms. Rebecca Morrison
American Finance & Mortgage
988-555-1236

"Nice resume," you say remembering the days when your own experience didn't amount to much. "Tell me about your job for American Finance and Mortgage."

"It was your basic clerical job," Carrie says. "Typing, filing, answering phones."

You nod, hoping she will go on. She doesn't. "And what does the Student Research Association do?" you ask.

"Oh, we get together every month and share new research techniques," Carrie said, seeming to brighten up a bit. "Mostly we talk about new Internet search engines, research sites, and things like that."

"Interesting," you say. "What research projects have you completed?"

Carrie was obviously prepared for this question. "I'm an English major, so I've done literary research. Some of the best work I've done was for my conservation classes. I recently finished a paper on the changing laws governing fur harvesting in the upper Midwest and how the laws have affected fur bearing animal populations."

You see a potential match between her experience and your needs, so you ask more particular questions. "How much time would you have, on a weekly basis, to conduct research for Donnelly Engineering?" you ask.

"About twenty hours a week," she replies. "Except for midterm and exam weeks, of course."

"Of course," you say. "School comes first."

You finish interviewing Ms. Smith and ask the next intern, Scott Harrison, to come into your office. You are immediately struck by Harrison's appearance—he is wearing a white shirt, navy blue sport coat, a red tie, and *jeans.*

"Hello," he says, vigorously shaking your hand.

"Nice to meet you, Scott," you reply. "Have a seat."

Scott doesn't wait for you to start asking questions. He leans forward in his seat, looks you straight in the eye, and says "I really want this job."

"Why?" you ask, slightly amused at his eagerness.

"Because I'm going crazy at school," he replies. "It's all theory, no practice. I want to get my feet wet, my hands dirty, you know?"

"I don't plan on sending you into a swamp to do research, but if that's what you want . . . " you joke.

"Whatever!" Scott exclaims, laughing. "Just put me to work!"

"May I see your resume?" you ask.

"Oh," Scott says a little more quietly. "I'm sorry. I didn't have time to prepare one after Dr. Jones had me call you. We just finished midterms. But I'll happily mail you one right away."

You are disappointed by his lack of preparation but decide to continue anyway. Before you can start, Scott continues. "Actually, my resume wouldn't be of much use to you. I haven't had any real jobs yet. That's why I'm so eager to start on something like this."

"What do you mean by 'no real jobs'?" you ask.

"Well," he replies, "I've worked on my father's farm my whole life. My family has been breeding and training quarter horses for three generations."

"Actually," you say, "that sounds like a very demanding job."

"Physically, yes," Scott says. "I sleep more in college than I ever did at home. But it did give me a pretty good work ethic. I know how to keep trying and trying until I get the results I want."

"What is your major?" you ask.

"Ecology," he answers. "I considered going for a preveterinary degree, but my sister is already a vet. I'm only a sophomore, but I've done well in my one hundred–level ecology courses. I have a 3.5 grade point average right now."

"Good," you answer. "What is it about ecology that interests you?"

"Everything," Scott says enthusiastically. "Tracing causes and effects, seeing the interconnectedness of natural systems . . . I just did a paper on the variety of effects the Chernobyl nuclear accident in Russia will have for the next twenty years. It was fifteen pages long, and I could have kept going!"

"I bet that would make a good senior thesis project," you say. "If you were to work for Donnelly, how much time per week could you devote to us?"

"Twenty, thirty hours a week," Scott says without hesitation. "I still wake up at six in the morning out of habit. I get most of my studying done before nine, because the dorm is quiet in the morning."

You finish the interview, and then ponder the striking differences between these two candidates. Jack mentioned that he wanted you to decide whom to hire, but he also wanted to reserve final approval. You make your decision as you turn on your computer to write up your thoughts and conclusions for Jack.

Suggested Next Steps

- Make a list of all the positive characteristics Carrie and Scott possess. Then make a list of the potential drawbacks to hiring Scott and Carrie.
- Decide which candidate you will hire.
- Using your list of characteristics and drawbacks, write to Jack defending your hiring decision.

CHAPTER 8

Short and Long Reports

As we have seen in previous chapters, employees in business and industry today write a variety of report styles and types. All of the strategies covered so far in this text—technical description and definition, instruction writing, business correspondence—are relevant to both the short and long report. Reports are one of the best workplace tools available for compiling, synthesizing, and presenting information pertaining to:

- recommendations
- project updates
- problem assessment
- evaluations

The primary difference between short and long reports, aside from the obvious implied length, is purpose. As you should be aware by now, the purpose and audience for any document has much to do with the type and complexity of the document. The same is true for the two report types. Short or long reports have distinct purposes and characteristics.

SHORT REPORTS

Short reports, which usually convey a brief run-down of activities, explain a problem, or describe an evaluation or assessment activity, tend to be brief (two to five pages), take a memo or letter format, and have less detailed explanations than do long reports. Short reports, therefore, are used in business and industry with more regularity than long reports because they are briefer, more encapsulated versions of activities or events. Short reports are also more manageable for employees with time constraints and deadlines. Remember to use a letter format for external short reports (reports being sent to someone outside of the company) and a memo format for internal short reports (reports being sent to someone within the company). Short reports, and to a lesser extent long reports, are used for distinct purposes and thus fall into distinct categories.

Types of Short Reports

Short reports are used for a variety of purposes and, therefore, are of different types. Short report types include action items, recommendations, project updates, problem assessment, and evaluations.

Recommendation Reports

Many reports are recommendations. The purpose of a recommendation is to persuade someone to follow your suggestions on a given matter. Recommendation reports focus on your belief—based on gathered facts and information—that a certain course of action be taken. For example, imagine that you are an assistant personnel manager in a small construction company. Your boss, the personnel manager, acting on a request from his supervisor, asks you to do a little research into the overall cost, repair record, and general reliability of three different copy machines. He wants a short report from you detailing the requested information and recommending which copier you believe is best suited for the needs of the company.

TIP

Use a recommendation report whenever you are making specific suggestions that will have an effect on colleagues, the workplace, general procedural requirements, or manufacturing. Remember that recommendation reports are based on fact-finding, information retrieval, and research.

Project Update Reports

Often called "progress reports," project updates are short reports designed to inform either a colleague or superior of your activities and advancement toward specific goals. Because projects are usually predicated on a specific timeline, updates, in the form of short reports, provide those individuals tracking your progress some sense of your movement toward completion. For example, imagine that you work as an advertising executive for a public relations firm. Your supervisor, who senses your potential, gives you a major ad campaign to devise for a new and innovative computer software company. You are held to a timeline, however, because the company plans to introduce a revolutionary desktop publishing program. You believe that three months is adequate time for developing the campaign, but your supervisor wants a project update every four weeks to follow your progress.

TIP

Use the project update report whenever you need to notify a colleague, client, or supervisor of your activities and advancement toward a goal. Remember, however, that the project update report varies depending on the project. These reports are most useful when you can provide not only a general sense of your movement but also specific dates by which certain tasks will be completed.

Problem Assessment Reports

Sometimes called a "problem analysis," the problem assessment report results from your study of a workplace problem or job-related situation and notifies a colleague, client, or supervisor of your findings. Sometimes a problem assessment report leads to specific suggestions, so the recommendation report and problem assessment do overlap in purpose in some instances. The problem assessment report is specifically designed for those situations that require research and study, findings, and analysis. For example, imagine that you are the assistant manager for a small manufacturing company. Your company has been very successful in producing and marketing a new line of office equipment specifically designed to reduce back problems and carpel-tunnel syndrome, conditions often reported by workers who spend more than five to six hours a day keyboarding. Your company has been less successful, however, in finding a freight company that can handle the increased demand, particularly because the equipment requires special care in shipping and your company has a two-day delivery guarantee. Your supervisor has asked you to study the problem by looking at rates, shipping time, weekend freight, and special contracts for exclusive shipping rights. You do not have to provide a specific recommendation on which company might win a shipping contract, but you must gather specific information about a variety of freight companies.

TIP

The problem assessment report does not presuppose recommendations but calls for objective analysis of data. Remember that if you are asked for specific suggestions or recommendations, you must provide adequate hard information to support your conclusions.

Evaluation Reports

Evaluation reports, as the name implies, are your assessments of the relative success or failure of a project, a campaign, a series of activities, or any job-related task with a specific goal. Because evaluation reports sometimes are about job performance or the abilities of others—colleagues, collaborators, support staff—you must gather information on their success or failure and discuss how they might improve or continue to succeed. For example, assume that you are the assistant line manager at an automotive plant. Your supervisor asks you to study and evaluate the relative success of a new line of robotic equipment. You must not only evaluate the usefulness of the new equipment, but you must also study and comment on the ability of the line staff to use and interact with the equipment. Your supervisor wants to see data indicating the amount of output, the productivity of the line staff with the new equipment, and the overall success or failure of the venture. Clearly, your supervisor wants your objective assessment of the new robotic equipment, but he also implicitly wants you to evaluate and comment on the general perceptions of the new equipment.

TIP

Evaluation reports require you to weigh carefully and consider the presentation of objective, quantitative data and information to support your subjective, qualitative opinion. Evaluation reports are sometimes difficult because the news you are reporting is not always positive. Remember, though, that if you present a balanced evaluation of the situation and outcomes, your opinion is likely to be supportable and valid.

Characteristics of Short Reports

Though short reports certainly vary according to intention, the style and format of the short report generally remains consistent. The most important considerations in planning and drafting a short report are

- audience
- style
- features

Careful consideration of these will likely result in a short report that meets your goals and communicates your message.

Audience

Audience considerations are always important regardless of the document you are drafting because you not only hope to articulate your thoughts in an appropriate manner, but you also seek to settle on a format appropriate to those readers. Consider these two important points when planning a short report.

1. When your short report (recommendation, project update, problem assessment, or evaluation) remains inside the company, use a memo report format. Refer to Chapter 2 for discussion of the memo report format.
2. When your short report is to be sent outside the company, use a letter report format. Refer, again, to Chapter 2 for information on letter style.

Remember that both memo and letter report formats are typically longer than simple pieces of correspondence. Both should contain, regardless of subject matter, a *report title* following the subject line in a memo and following the inside address in letters. Refer to the sample short reports (Figures 8.1 and 8.2 on pages 207–213) for a model of both memo and letter report format.

Style

Short reports have a distinctive style because they tend to be brief and are used to communicate specific information. Follow these general style guidelines when writing a short report.

1. Use headings to separate main ideas or to group important information. Readers will find your report easier to follow with headings that guide the movement and progress of your document.
2. Bullets or numbered lists attract attention, so use them when you want to direct readers' attention to short lists of information or numbered steps.
3. Leave adequate white space so that your report is not congested. Remember that memos and letters are usually single-spaced, so be sure to leave at least two spaces between paragraphs and possibly three spaces between major sections with a heading.
4. Write in the active voice and move from general to specific information both in the overall format of your report and in the individual paragraphs themselves. Readers tend to understand your message more easily when they are first presented with a general, broad point and then a series of specific, concrete facts.

These general guidelines are designed to help you write short reports that will appeal to your audience while also communicating your essential message. In addition to specific style guidelines, consider the following special features unique to short reports.

Features

The special features or characteristics of short reports typically remain consistent whether you choose memo or letter format. Because a short report has as its goal the communication of important, usually factual information in a relatively brief format, adherence to these features will insure a more consistent document. The short report is made up of three parts: the executive summary, the body, and the conclusion.

Executive Summary. The executive summary, sometimes called an "overview," either follows a heading of the same title or comes directly after the subject line in a memorandum report or the greeting in a letter report. Either way, the executive summary provides the reader an abstracted version of the whole report, a condensed view of what you plan to present in the body. The executive summary is a popular feature of short reports for busy people or those who seek primarily your "bottom line" point. The executive summary, therefore, should not be lengthy (one or two paragraphs) and should present a broad description of the contents of the report. Refer to the sample short reports on pages 207–213 for an example.

Body. The body portion of the short report, as well as the heading that introduces it, varies depending on the report type. For example, in a *problem assessment* report, the body portion might be introduced with the heading *Problem* or *Problem Assessment*. The heading for an *evaluation report* might use the heading *Evaluation,* and the heading for the *project update* would likely be either *Update* or *Progress Report*. Regardless of the heading you use to introduce this important portion of the report, the body contains the useful, detailed information your audience will rely on to make a decision or take action.

TIP

If you are writing a short report in response to a specific request, use language consistent with that request. For example, if your employer asks you to review budgetary problems in your department, use a heading like *budgetary status* to reflect your attention to detail.

Again, refer to the sample short reports on pages 207–213 for examples of the body portion of a short report.

Conclusion. As the name implies, the conclusion gives you an opportunity to wrap up your comments and, in some cases, to provide overall recommendations (if you have not done so previously in the report. Use the heading *Conclusion* to introduce this section. Typically a summation, the conclusion reminds readers of your main points, your suggestions or evaluations, and your "bottom line" point. The conclusion, therefore, should not be lengthy (one or two paragraphs).

The goal in writing a short report, whether it be an internal or external document, is to provide the reader important information and usually suggestions, evaluations, or recommendations based on that information. Sometimes, however, you may be called on to produce a report that requires more research, more detail, and more support in the form of attachments or appendices. Such a report is the *Long* or *Formal Report*.

EVALUATION OF POTENTIAL
YOUTH RECREATION FACILITY SITES

Executive summary

Executive Summary

Alternatives for Area Youth (AAY) performed a limited study of possible locations for a youth center. AAY considered only centrally located Newton area sites for several reasons. First, youth are already congregating in downtown Newton and therefore obviously have found the means to transport themselves to this area. Second, young people do not have ready access to their own transportation, so requiring youth to walk or bike to a recreation facility is necessary. The downtown area is well-lit, and regularly patrolled by Newton city police officers and is a safe place for a youth center. After careful consideration of each site, AAY has no specific recommendation; however, the group felt using the former Club Newton building is the most cost-effective and expedient solution to opening a youth facility. A discussion of the pros and cons of each site identified follows.

FIGURE 8.1 Sample Short Evaluation Report

Evaluation report body

The Former "Friends" Restaurant

This site, a former restaurant located next to the New Music Center on Case Street, offers a centralized location and has two to three floors available for various uses, such as a dance hall, a small pizza and soda shop, game room (video and pool), and study areas.

The owner of the building, however, is asking $800 per month in rent with an increase of $100 per month for each year of occupancy (i.e., $800 per month for 1999, $900 per month for 2000, $1000 per month for 2001). AAY feels this amount would exceed a youth center rent budget and could not compromise with the owner on the rent. Furthermore, as the site previously housed a bar, AAY feels this location's reputation may cause concern among parents.

Vacant Case Street Commercial and Residential Building

Another Case Street site, this one a commercial and residential building on the corner of Case and Hill streets, which has sat vacant for the last six months, presents a more "home-like" atmosphere and offers easy drop-off and pick-up on the side street. By purchasing this house (the asking price is $65,000), the youth center could avoid problems with leasing (e.g., landlord concerns over damage to the facility). Moreover, this purchase would add to the city's assets and may have other uses should a youth center fail to succeed.

The difficulties associated with purchasing this house include:

- securing a mortgage
- finding additional funds for extensive remodeling
- Assuming an annual property tax burden of approximately $900 per year
- Keeping noise from dances and other activities to a minimum in consideration of residents within 50 feet of the site

Additionally, Newton's Youth for Christ organization is considering this site for their own operations, so the commission

FIGURE 8.1 Sample Short Evaluation Report *continued*

would need to move quickly if interested in purchasing the building.

The Value Hardware Building

The former Value Hardware building located on Main Street offers an extremely large area that could be remodeled to suit a variety of purposes. Additionally, the site's central location and the nearby city parking lot make it a feasible site for a youth center.

AAY feels that, although the Value Hardware building offers more physical space than any other available location, it consists of large, open spaces and would require extensive remodeling to provide separate areas for different activities and staff offices. Additionally, this building is located next to Harry's Bar, and thus would not provide the atmosphere for a safe, alcohol-free youth hangout. The proximity to this rather notoriously rowdy bar will undoubtedly cause parental concern.

Club Newton

AAY seriously evaluated the possibility of reopening the Club Newton site. This building contains three levels and already has a dance floor. Club Newton's location on Main Street is a well-lit and well-known area. Additionally, the former club's organization still has control of funds from its previous operation and is willing to turn those funds, $2,000, over to the city to facilitate a speedy start-up. The location of the club presents problems. During the Club's operation, several young people were injured while crossing busy Main Street to "hang out" in nearby Main Street Park. The club gained a poor reputation because of the activity in the park—young people were rumored to have been drinking and smoking in the park, then returning to the dance (where staff strictly prohibited any such activities).

Conclusions and Recommendations

Conclusions and recommendations

AAY understands the final site decision rests with the Commission as it evaluates location, traffic, parking, and budget issues involved in establishing a permanent youth facility. If low start-

FIGURE 8.1 Sample Short Evaluation Report *continued*

up costs become a deciding factor, AAY feels the former Club Newton site has the most to offer.

AAY, however, has learned through researching the failure of Club Newton, that any facility should adopt a strict policy prohibiting youth from leaving and reentering during activities. Furthermore, there should be a concerted effort to identify young people who arrive intoxicated or high, and hold them until either their parents or the authorities arrive to remove them from the facility. AAY feels that wherever a new youth facility is located, its staff will have to "prove" their commitment to a substance-free atmosphere to alleviate current parental concerns.

FIGURE 8.1 Sample Short Evaluation Report *continued*

Executive summary

Executive Summary

Alternatives for Area Youth (AAY) has conducted an in-depth analysis of the causes of the July 1998 protests by Newton area teens. While these protests may appear to be in reaction to tickets issued by law enforcement (in accordance with new city ordinances) for bicycling, in-line skating, and skateboarding on downtown sidewalks, AAY believes several other factors are involved.

During the summer months, the majority of teen recreational activity in the Newton area centers on socializing—gathering in large groups to talk, eat, and exhibit talents in bicycling, in-line skating, and skateboarding. When organized activities are offered by the city or various youth-oriented groups, they are well attended (as evidenced by the popularity of the former Club Newton). However, these events are sporadic and teens will invariably resort to their least expensive recreational option, "hanging out." This particular activity has become more difficult in recent months because of several issues including

- the closing of Club Newton, a popular gathering place that sponsored dances and youth-oriented activities

FIGURE 8.2 Sample Short Problem Assessment Report

- the implementation of policies by local merchants that prohibit young people from using fast-food and other restaurants as gathering places
- Enforcement of so-called "sidewalk ordinances" that result in high fines for bicycling, in-line skating, or skateboarding on any downtown street at any time

Problem assessment report body

The Closure of Club Newton

Newton's only teen night club closed in the spring of 1998. The club, started with grant funds, suffered from poor management and planning. Although some funds still remain in its accounts, disagreements between the parental and teen committees made it impossible to conduct day-to-day operations. In addition, poor staff training made it possible for teens to arrive intoxicated and remain on the premises, which eventually damaged the club's reputation for substance-free programs.

Despite its problems, the club regularly enjoyed a large attendance at its after-school and evening programs. According to former staff members, over a hundred teens would regularly come to the club after school and during the evenings to buy snacks, relax, play pool, and talk in one of the many "hangout spots" the club provided by moving chairs, tables, and couches into groups on the large dance floor. The club also provided quiet study rooms and tutoring by volunteer college students. During weekend dances, the club regularly drew crowds of over two hundred teens. The club's closing, therefore, left a significant void in youth recreation opportunities.

Implementation of Prohibitive Rules Against Young Patrons

Another "place to go" for young people is an inexpensive restaurant to eat and talk. However, a few incidents with teenage patrons (e.g., loudness and creating a mess) resulted in an over-reaction by area merchants, particularly restaurants, in the form of prohibitive policies. Many fast-food restaurants hired guards for evening and weekend shifts to enforce new rules, such as not allowing teens to sit and talk after they finished eating (regardless of their behavior). Another restaurant on Main Street demands that each teen purchase at least five dollars worth of food (not including beverages); otherwise, they are asked to leave. Adults who order coffee and dessert, however,

FIGURE 8.2 Sample Short Problem Assessment Report *continued*

are never subjected to this policy. As a result of their inability to "hang out" for a few hours in a restaurant or fast-food place, even more teens took to wandering the streets of downtown Newton.

Enforcement of New "Sidewalk Ordinances"

In June of 1998, the Newton City Police Department Chief appeared before the City Commission to request a new ordinance prohibiting the use of skateboards, bicycles, and in-line skates on downtown area sidewalks. The police chief cited several complaints of local merchants that teens gather in city-owned parking lots near their businesses during late afternoon and evening to watch other teens "perform" on bikes, skateboards, or in-line skates. Two incidents of collisions between pedestrians and teen skateboarders had occurred, in addition to one car–bike collision. Citing their concern for both the teens and people walking in the downtown area, the City Commission passed a "sidewalk ordinance" that made it illegal to use any wheeled vehicle or piece of sporting equipment on downtown sidewalks or in city parking lots. The fine for the first infraction is $50, with second offenses ranging from $200 to $500.

On July 14, 1998, police officers began foot patrols of Case and Main streets, issuing fifteen tickets for violating the new ordinance. All tickets were issued to teenagers. The officers reported averaging five or six tickets per night for the next several days, during which the teens protested by picketing City Hall. According to police reports, seven downtown businesses were vandalized between July 14 and July 20 (the suspects were teenagers), and a great deal of graffiti appeared on the walls of the city-owned parking garage. During the same period last year, Newton police records indicate no incidents of vandalism, and no orders to remove graffiti on any city properties were issued during the entire month of July 1997.

Conclusion

Conclusion

The protests of July 1998 did not result solely as a reaction to the "sidewalk ordinance." Police concern over the parking-lot gatherings was valid, considering the sudden rise in the numbers of teens participating in this kind of activity.

FIGURE 8.2 Sample Short Problem Assessment Report *continued*

The rise in teens using the streets of Newton as their recreational grounds is not, however, surprising given the loss of Club Newton and the new restrictions on patronizing fast-food and other restaurants. Teens continue to gather in city-owned parking lots, and merchants have recently made new complaints to police and city officials about loud music from car radios and suspected drug and alcohol usage occurring near their businesses.

FIGURE 8.2 Sample Short Problem Assessment Report *continued*

LONG OR FORMAL REPORTS

Compared to the short reports, long reports do not really fall into types or categories of reports. Long reports are distinctive because they can cover virtually any topic but are usually produced and formatted in a consistent pattern regardless of topic. Long reports are the choice of writers who need to present a more complete package of information to their supervisors, colleagues, or clients. Such reports, because they often involve attachments, usually run longer than the five pages, the maximum length (typically) of the short report.

Long reports often have a variety of readers with different levels of technical background. As such, different readers will rely on different portions of your report for the information they seek. For example, if you write a long marketing report on three different computers your company might consider purchasing, your supervisor is likely to be most interested in your "bottom line" recommendation (found in either the executive summary or at the conclusion of the report). The technical support staff, responsible for installation of the new computers, is more likely to read portions of the body and appendices that detail special features of the computers. Coworkers, who will be the actual users of the new computers, will probably scan the table of contents to search out information on the "user friendliness" of the new equipment. Thus, regardless of the topic of the long report, expect a variety of people to read portions of your work as they search for the information most relevant to them.

Characteristics of Long Reports

Though long reports vary according to intention, the style and format of the long report remains generally consistent. The most important components of a long report are:

- style
- features

Careful consideration of these two features will likely result in a long report that meets your goals and communicates your message.

Style

Long reports have a distinctive style not only of characteristics but also of the specific information they communicate. Follow these general style guidelines when you are drafting a long report.

1. Use headings to separate main ideas or to group important information. Readers will find your report easier to follow with headings that guide the movement and the progress of your document.
2. Use bullets or numbers to set off lists of points or important information. Bullets or numbered lists attract attention, so use them when you want to direct readers' attention to short lists of information or numbered steps.
3. Leave adequate white space so that your report is not congested. Long reports can be either single or double spaced, but make sure the spacing between paragraphs is consistent. Leave adequate space between major sections with headings.
4. Write in the active voice and move from general to specific information both in the overall format of your long report and in the individual paragraphs themselves. Readers tend to understand your message more easily when they are first presented with a general, broad point and then a series of specific, concrete facts.

These general guidelines are designed to help you write a long report that will appeal to your audience and communicate your essential message. In addition to specific aspects of style, consider the following features unique to long reports.

Features

Adherence to the following special features or characteristics of long reports will insure a more consistent document. Long reports are usually made up of seven distinct parts, not including the appendices and a table of illustrations (both optional). Those parts, in the order they will appear in the report, are:

- title page
- letter of transmittal
- table of contents
- executive summary or abstract
- overview
- body sections
- conclusion

Title Page

The title page, which immediately follows the cover of your report, contains your name, the date, the title of your report, and sometimes the person for whom the report is intended. Refer to the sample long report on page 219 for an example of the title page.

Letter of Transmittal

The letter of transmittal is unique to the long report. Because short reports are already correspondence-type reports (memos or letters), the long report needs a letter to indicate for whom the report is intended and to preview the contents of the report. Use the standard conventions of letter formatting in Chapter 2 as you write the letter of transmittal. Use the letter to preview the long report, to hint at possible recommendations and to establish the framework of the material that follows. Refer to the long report on pages 220–221 for an example of the letter of transmittal.

Table of Contents

All long reports need a table of contents, a convenient tool for readers who seek quick information and a glance at the highlights of your work. The table of contents, which directly follows the letter of transmittal, should be arranged according to the major headings you have used in the report body. Use the exact wording in the table of contents so readers will readily make connections from it to the body of the report. In addition, follow these guidelines:

- *Parallelism.* Make both the headings in the body of your report and the headings reflected in the table of contents parallel. For example, if you begin using headings that follow an "ing" pattern (*Estimating the Cost, Determining the Feasibility,* and *Excavating the Site*) do not switch to a different heading style (*Cost to Investors* or *When to Begin Digging*) because this can be confusing to readers.
- *Page Numbers.* Use consecutive page numbers in the long report, and use those numbers in the table of contents.
- *Boldface Type.* Use bold face type or underlining to indicate major breaks in the report body and in the table of contents. Use adequate white space in the table of contents so the page is not congested.
- *Attachments.* List, in the table of contents, all attachments or appendices. Give these attachments either a page number or an appendix letter (e.g., Appendix A) for quick reference.

Refer to the sample long report on page 222 for a sample table of contents.

Executive Summary or Abstract

Just as you learned to include a summary in your short report, so you should include one in the long report as well. While both the executive summary and the abstract condense and summarize information, they are written for different purposes. If the long report is for an executive or manager, use the executive summary. If the report contains the results of a scientific or technical inquiry, use an abstract. The long report summary is distinct from the short report summary in that (a) you are using more detailed material and (b) the reader needs a more developed summary. As such, the executive summary and the abstract tend to be longer but should not exceed a single sheet of paper. Because the summary (in either form) is separate from the report text, do not follow the summary with the report introduction. The summary condenses and presents the primary points in your report, so emphasize, in particular, the recommendations you will make. Refer to the sample long report on page 223 for an example of the summary.

Overview

The actual report opens with the overview, an introduction to your long report that emphasizes the overall scope of your project and the primary points you will cover in the body of the report. The overview, thus, is not lengthy (usually one page or less) and forecasts for the reader the important sections to follow. Refer to the sample paper on pages 223–224 for an example of the overview.

Body

Though different in both length and detail, the body portions of the long report are similar to those of the short report in that both provide the reader with details that support your main point or ultimate recommendations. The body portion, thus, has major headings (main points) and subheadings (minor or supporting points). Headings are reflective of the primary information, are parallel in structure and are complete units. As such, headings vary greatly from report to report. To make the body paragraphs as successful as possible, follow these guidelines.

- Move from general to particular as you write the paragraphs. Offer the reader a general point and follow that point with specifics.
- Use bulleted or numbered lists to present information that you want to make sure the reader catches.
- Use adequate white space in the body portion. A dense text can cause the reader to skip over important information.

Refer to the sample long report on pages 224–229 for an example of the body portion of the long report.

Conclusion

The conclusion of the long report, like that of the short report, provides readers your overall recommendations and suggestions. The conclusion is typically longer and more thorough than that of the short report. Include a detailed list of your recommendations and suggestions or your evaluation of a problem. Refer to the sample on page 229 for an example of the long report conclusion.

Optional Features: Table of Illustrations and Appendices

In some long reports, you may need either a table of illustrations or appendices. Because both are optional, follow these general guidelines when trying to decide whether or not to include this material.

- *Table of Illustrations.* Include a table of illustrations, immediately following the table of contents, when you have four or more visuals, figures, or tables in the report body. Format the table of illustrations in the same manner as the table of contents, using figure or table numbers to denote individual visuals or tables. Refer to the Guide to Creating and Working with Visuals for more information on using visuals and placing them in texts.
- *Appendices.* Appendices are usually included at the end of a long report, immediately following the report conclusion, when you have material that is not appropriate or too lengthy or detailed to include in the report body but important enough you would like your readers to have access to it. For example, a long report on the feasibility of a new computer lab in a local high school might include relevant information on cost and site selection in the report body. A blueprint of the new lab and detailed cost breakdowns of equipment and furniture, however, would be too exhaustive for the report text and would be more appropriate as a series of appendices.

Appendices are typically used by readers seeking specific or specialized information that might not be relevant to a more general readership. As a rule of thumb, though, restrict appendices to no more than six additional pages of material. A report with too many appendices can seem overwhelming to many readers.

Regardless of the report type or report length you decide to write, remember always to consider the needs of your audience, to write coherently and succinctly, and to use the active voice. The classroom exercises will give you practice writing short reports and thinking about the components of the long report.

Title page

Proposed Plan for Improving Youth Recreation Opportunities

Respectfully Submitted to the Newton City Commission

September 27, 1999

By

Alternatives for Area Youth

FIGURE 8.3 Sample Long Report

Letter of transmittal

Alternatives for Area Youth
P.O. Box 168
Newton, MI 55555

December 13, 1999

Newton City Commission
Newton City Hall
Newton, MI 55555

Dear Board of Commissioners:

Alternatives for Area Youth (AAY) addressed the Commission on September 27, 1999, concerning our ongoing youth recreation study.

AAY documented the following problems stemming from Newton's few youth recreational opportunities:

- young people congregating on city streets, causing traffic hazards
- friction between youths and merchants, residents, and law enforcement officials
- Incidents of vandalism to businesses, homes, and private property

AAY proposes various solutions to alleviate these problems, some for the short term and others that require long-term planning and the identification of significant funding sources. These solutions include

- opening all city-owned facilities to youth programming free of charge and coordinating regular activities in these facilities
- designating an area in City Park as a "Summer Youth Park" and allowing young people to congregate there under park employee supervision
- conduct forums with area businesses in an effort to form a protectionist, rather than antagonistic, attitude toward their young patrons
- Begin a process, outlined in this report, to gather community-wide support for the creation of a permanent youth facility

FIGURE 8.3 Sample Long Report *continued*

AAY trusts this Commission understands the seriousness of this issue. Our members are available to answer any questions the Commission may have concerning this report.

Sincerely,

Jan Smith   Shawn Lawson

Jan Smith, AAY Chair   Shawn Lawson, AAY Vice-Chair

FIGURE 8.3 Sample Long Report *continued*

Table of contents

TABLE OF CONTENTS

FIGURE 8.3 Sample Long Report *continued*

Executive summary

RECOMMENDATIONS FOR NEWTON CITY YOUTH RECREATION PLANS

Executive Summary

Alternatives for Area Youth (AAY), an organization dedicated to facilitating improved recreation options for Newton City young people, recommends creating a location and programs for area youth. We believe that providing a location for teens to congregate will alleviate friction among these young people and the merchants, residents, and law enforcement officials of Newton.

As indicated in our previous report detailing the various problems experienced by the Newton community, among the possible solutions is a youth recreation center. This report recommends plans for and alternatives to starting a youth center, including staffing structure, activities, and programming.

In conclusion, we recommend that the Newton City Commission coordinate efforts by the Brighton Foundation, the remaining members of the former Club Newton staff, Newton Youth for Christ, Newton Community Housing, and other youth-oriented organizations to avoid duplication of services and to meet the diverse needs of area youth. We also recommend that the Newton City Commission make youth recreation part of its long-term budget plans. Newton must take a protectionist, rather than antagonistic, view toward its young people.

Overview

Synopsis of Problem

Many Newton residents believe the community's "youth problem" suddenly appeared during the summer of 1998, when police and young people clashed in and around the Case Street area. After researching the issues, however, AAY discovered the summer's discord was a result of a pattern of half-hearted attempts to provide young people with constructive recreational activities. Some of the contributing factors to the conflicts between youth and the residents, merchants, and police of Newton were the closing of Club Newton and the implementation of restrictive policies by restaurants and fast-food

1

FIGURE 8.3 Sample Long Report *continued*

places. With no "place of their own" and no other safe indoor area to meet and talk, teens wandered the downtown streets and congregated in parking lots. The traffic congestion, noise, and occasional accidents between pedestrians, cars, bicycles, skateboarders, and rollerbladers naturally attracted the attention of law enforcement. New ordinances such as a ban on cycling, skateboarding, and in-line skating in downtown areas further frustrated teens and resulted in the angry protests staged on Case Street during July 1998.

Long report body

Potential Effect of Youth Recreation Facility

All groups affected (youth, parents, merchants, law enforcement) benefit from the establishment of a facility (youth center) and activities (programs) for area youth. Likewise, if nothing occurs, all groups involved stand to lose more as divisions between teens and the rest of the community deepen. When young people are given a "place of their own," the community experiences fewer conflicts, as evidenced by the significantly lower number of youth congregating on the street when programs such as Club Newton were in place. Other communities, such as Crystal, Michigan, which has had a youth center for four years, experienced a sharp decline in vandalism and underage drinking immediately after their center opened.

Recommended Youth Programs

During the fall of 1998, AAY surveyed 51 high school students to determine their desire for a youth center. In addition to questions about the center's location and operating hours, AAY asked several questions about youth programming. The majority of these young people wanted dances and pool tables (the two major activities and equipment at the former Club Newton). As for after-school programs, the survey results indicated a need for comfortable places to study, to receive tutoring, to listen to music, to use computers, and to talk, eat, and play video or other games such as pinball and foozeball. Several students asked for center-sponsored events such as volleyball or basketball tournaments, trips to larger cities during school vacations, and various contests to win prizes donated by local merchants.

2

FIGURE 8.3 Sample Long Report *continued*

All these programs appear worthwhile; however, AAY believe the young people themselves should determine what programs occur and when. The creation of a programming board, discussed later in this report, is essential to offering programs that interest young people, while giving them an opportunity to learn new skills in the planning of those programs.

Recommended Facility Operation Plan

The youth center requires staffing on two levels, including two administrative boards and an operational staff. The executive director coordinates these groups. AAY suggests the following organizational structure:

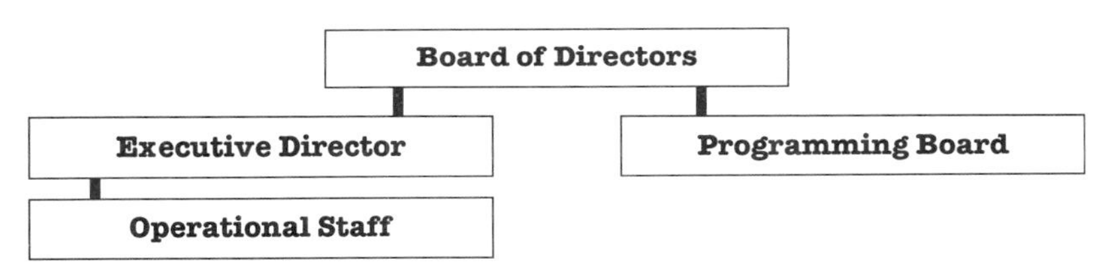

Figure 1: Recommended Organization Structure

Administration

The administrative portion of the organizations consists of the executive director, the board of directors, and the programming board. The board of directors hires the executive director and approves the members of the programming board. The programming board hires the operational staff.

The Board of Directors

This board consists of five people: two area merchants, two citizens-at-large, and one person under the age of eighteen. A member of the City Commission serves as the board's chair. The responsibilities of the board of directors includes

- selection of an executive director
- approval of programming board members
- creation of yearly budgets

3

FIGURE 8.3 Sample Long Report *continued*

- location of funding sources
- support of the center's programs

The City Commission will appoint three members to the board of directors. The Newton City High School student council will select one citizen-at-large and the under-eighteen board member.

The Executive Director

The executive director takes charge of day-to-day operations. He or she answers to the board of directors and oversees the operations staff while working closely with the programming board. The executive director's salary is paid from the facility's operating budget. The executive director is also responsible for the following:

- hiring staff
- recruiting volunteers for programs
- justifying all expenditures
- overseeing public relations
- scheduling events and appropriate staffing
- preparing and delivering progress reports to the board of directors

The executive director must have a strong ability to work with young people. AAY recommends hiring someone with a background in education and experience in public administration. The director holds the only full-time position within the organization.

The Programming Board

This essential part of the center's administration consists of six students: four from Newton Senior High School and one from each Newton middle school. These students are chosen by their schools' student councils. Each member of the board chooses the type of activity he or she wishes to organize and a simple majority vote of the programming board approves the activity. The executive director organizes these meetings and assists board members with implementing their programs.

4

FIGURE 8.3 Sample Long Report *continued*

Operational Staff

The operational staff consists of a mixture of volunteers and part-time paid staff. The operational staff implements the activities approved by the programming board. The executive director oversees the recruiting of volunteers and staff and conducts appropriate staff training. AAY recommends that the director draw upon existing programs and recruit volunteers from these programs to become either directly involved in the center or in staff training. For example, the door tenders at center dances should be trained in how to identify and handle intoxicated young people. The resident assistants at Superior University receive such training and are, therefore, a perfect resource. All staff members should complete first aid and CPR training available through Newton's chapter of the Red Cross.

Alternatives to recommended actions

Alternatives to Creating a New Youth Facility

In the event that the City Commission decides against the creation of a youth center, AAY recommends several other courses of action. These activities are not without cost to the city in terms of staffing facilities, and should become part of the city's regular budget.

Utilize Existing Parks and Recreation Department for Youth Activities

One of the city's greatest assets are its many parks, which can serve youth needs in both summer and winter. AAY recommends the following courses of action to provide recreation for young people.

Creating a "Summer Youth Park" in a Portion of City Park

According to AAY's youth survey, one of the places at which young people enjoy gathering is City Park. The park has a large picnic area, a beach complete with volleyball net, a baseball field, basketball court, and restroom and shower facilities. AAY recommends that the city designate the north end of the park as a "youth park" and allow young people to congregate there under park employee supervision.

5

FIGURE 8.3 Sample Long Report *continued*

While there is no policy concerning how many young people are allowed to gather at the park, the city's employees regularly chase off teens when more than twenty or so congregate. When asked why the park discouraged young people from using its facilities, Mr. Smith (director, Parks and Recreation Department) commented that he was concerned "the kids were using drugs." No confirmed incidents of drug use have occurred, however, and AAY feels this concern is without merit. The north end of the park is closest to the Parks and Recreation office, where employees can easily monitor youth activity. AAY strongly suggests the city makes it clear to the Department of Parks and Recreation that young people are welcome in the park as long as they abide by the posted park rules concerning noise, alcohol, and vandalism.

Improve South Beach

South Beach, located within walking distance of downtown Newton, is another favorite summer teen hangout. AAY recommends hiring a lifeguard for the summer and constructing a volleyball net on the beach. The presence of a lifeguard will alleviate parents' concerns for their children's safety, and the volleyball net will attract more teens as beach volleyball is very popular.

Open the High Street Gymnasium for Youth Activities

The city owns a large building on High Street, part of which is used for a Senior Citizens Center. The building also contains a gym and is perfect for an after-school basketball program. The Zcity could also sponsor youth dances and small concerts (such as a coffee-house-style performance by local bands) in this facility.

Create Ice Rinks in City Park

AAY discovered that the Parks and Recreation Department used to flood the lower end of City Park's baseball field during winter, creating a large ice rink. This program was discontinued in the 1980s when the person who created and maintained the rink retired from the department. AAY recommends resur-

6

FIGURE 8.3 Sample Long Report *continued*

recting the winter ice rink and promoting it among teens as a place to organize their own hockey games or simply to come and skate. The city could recoup some of the cost of the rink's creation and maintenance by selling a vending contract for the sale of hot beverages in the evenings and on weekends.

Conclusion

Conclusion

While AAY believes a permanent youth facility represents a long-term solution to the friction between young people and the community, our organization also has provided the City Commission with several suggestions on how it may exhibit its commitment to the safety and positive development of our youth. Continuing to do nothing will only exacerbate the frustration hundreds of teens so clearly displayed during their organized protests this past summer. If our young people can get together to voice their feelings, surely the city can respond by organizing positive outlets for their energies. It will take time to establish a successful youth center, and AAY believes the city has many options to consider to ensure that tax dollars are well spent. In the meantime, AAY hopes the City Commission will implement some of the alternatives provided within this report while planning for a long-term commitment to a permanent youth-oriented facility.

7

FIGURE 8.3 Sample Long Report *continued*

Exercises for Classroom Discussion

1. You are a member of a student public relations team assigned to study and report on the cost and availability of housing for students considering transferring to your institution. Your assignment is to write a short report, with your recommendations, to the university's Associated Student Government (ASG). Is this an internal or external report? Does this topic involve more detail than you can include in a short report?

2. Write a project update report, to your instructor, detailing your progress in the class thus far. Report on your work, your writing, your ability to understand the material, and your plans for future work. Follow the format detailed in this chapter.

3. You are the Assistant Personnel Manager of a small construction company in the Midwest. Since you started working for the company, you have heard your supervisor, the Personnel Manager, complain about employee theft of small, inexpensive office items such as pens, paper, tape, and paper clips. Initially the thefts were overlooked because they didn't occur often and they involved relatively inexpensive items. Lately, though, the problem has grown, and the company is spending more and more to replenish its inventory of stationary items. Your supervisor has been charged by the company vice president to study the problem and produce a short report. She has, in turn, assigned you to write the short problem assessment report. Specifically she would like to know how the problem arose and why employees feel free to take such items.

4. You are a student member of a university group assigned to study tuition structures and evaluate three types of tuition alternatives: (a) a flat-rate fee for all students taking between twelve and eighteen credit hours, (b) a standard, per-credit fee (the university's current model), and (c) a per-credit fee for one to twelve credits that changes to a flat-rate fee when a student takes between thirteen and twenty credits. As the only student member, you are to write a short evaluation report summarizing your feelings about the three tuition plans and recommending the best one for the majority of students attending the university.

5. You are a computer lab assistant at the institution you now attend. Since you first enrolled for classes, you and your friends have complained—and have heard others complain—about the lack of lab space and the poor computer availability for full-time students. Ironically, just as you begin your new job at the computer lab, the university decides to study the computer problem, and your supervisor, the assistant director of computing services, has asked you to study the problem and report back to him on possible solutions to the problem. Specifically, your supervisor has asked you to analyze the problem, the student complaints, the available space for new labs, the potential cost of adding computers, and a possible redesign of the existing labs to add computers. He wants you, as well, to study more efficient means for making the labs available to students. This long report will take some time to research and put together.

SCENARIOS FOR SHORT AND LONG REPORTS

The Trouble with Traffic

CONSIDERATIONS

In this situation, you have decided to volunteer your skills to help prevent car–pedestrian accidents on your campus. As you read the following scenario, keep these questions in mind:

- What information is at your disposal to help you create this report?
- Who are your readers, and what are their potential attitudes and concerns toward this report?
- What results do you hope your document generates?

During your college years, you have had experiences (to add to your resume) such as volunteering to build houses for the homeless and marching in support of different causes. While advocating change for the homeless and being an activist were interesting, you want to do something that has a more lasting impact. So when your school's student government announced it would tackle a long-standing problem—the traffic flow in and around campus—you decided to volunteer your time and energies to the cause.

You realize that traffic flow would not seem to some to be worthy of too much attention, but you have your own reasons. Last year, your best friend was hit by a car while walking home from the library late at night. After leaving the hospital with a broken leg and collarbone, your friend spent months in physical therapy. You later found out that several car–pedestrian collisions had occurred on campus.

Today, you are meeting with Greg Borden, a student government representative in charge of the traffic problem initiative.

"So where do we start?" you ask after the two of you exchange "hellos" and discuss mutual friends.

"Well, we need to pinpoint the worst traffic problems on campus," Greg says. "I guess we could talk to the public safety office, and maybe the city keeps records on accidents that happen at the intersections between campus and city streets."

"What about just talking to people?" you say. "I mean, how many close calls have we all had at certain places? They don't show up on police reports."

"Good idea," Greg says. "For now, let's just do an informal survey. We don't have to get really fancy and print up questionnaires or anything."

"Okay," you say. "I'll get my hands on a good campus map, so we can mark all the potential trouble spots we find. By the way, what are we going to do with all this information?"

"Well, the first thing to do is to create a report for the next student government meeting," Greg answers. "Whatever we come up with in terms of solutions to the problems we find, they'll have to decide whether or not to pursue. Then it's up to them as to what we do next, but I would imagine they'd have us work with whomever has the power to change the roads, sidewalks, signs, and whatever else needs fixing."

"We have a lot of work to do," you say with a sigh.

"And we'd better be right on the money with what we say," Greg says. "The last time the student government asked for changes in the parking lots, we were told our concerns were unfounded. If there's any problem that we can't prove is really, truly, and without doubt a problem, then we're either leaving it out or getting more information on it."

After leaving Greg, you think about all the places you had had close calls, either while driving, walking, or riding your bike around campus. *What if we can't prove that those intersections are dangerous?* you ask yourself. You remember that a campus safety officer had mentioned complaints about the place where your friend was hurt. After that accident, the university replaced the YIELD sign with a STOP sign. *What's proof?* you think. *Someone getting seriously hurt?*

SUGGESTED NEXT STEPS

- Review the guidelines for creating reports in this chapter.
- Conduct the necessary research you need to gather information on your campus' traffic problems. If you have a large campus, you may choose a section to analyze. You should work with your classmates as you request information from campus safety or other law enforcement officials. Document your research techniques and information sources in your Solution Defense.

- Carefully consider the issue of "proof" for problems and discuss your definition of "proof" in your Solution Defense.

The High-End Real Estate Plan

CONSIDERATIONS

Marketers do much more than sell a product or service—the bulk of their work is determining the most effective methods of reaching consumers. In this situation, you are evaluating different advertising media and determining which publications best match your client's needs. As you read the scenario, keep the following questions in mind:

- What is being sold, and how does the nature of the product affect how it is marketed?
- What are the variables in choosing a publication within which to advertise?
- What outcomes do you desire from the creation of your document?

As a new member of the marketing team at Concepts, Inc., you are responsible for making recommendations to your clients concerning their advertising and marketing activities. Concepts, Inc. is a full-service marketing and advertising firm, offering assistance to businesses in planning and creating print, television, radio, and Internet advertising. Each Concepts, Inc. consultant handles five to seven clients, whose needs require at least 50 hours of work per month. Your biggest client is the Waterfront Realty, a company specializing in rather expensive lakeshore and recreational properties. This year, Waterfront has taken on the task of selling land in two areas. The first is Pebble Island, which is located in Lake Ontario about a five-minute boat ride from the nearest town. It is selling for $900,000 with no option to subdivide the property. The island has a rustic cabin, hand-pump well, and electricity generator but no indoor plumbing. The second development is a new subdivision, Agate Point, on the shores of Lake Ontario. Each two-acre Agate Point lot has either a view of Lake Ontario or a portion of the shoreline with prices ranging from $50,000 to $150,000. This development has paved roads, driveways, and electricity and is accessible all year.

The owner of Waterfront Realty, Mark Storey, has asked you to find several magazines whose readership would match the type of people who would buy north country waterfront property. He has an advertising budget of $40,000 and wants to spend approximately half of that on magazine advertisements.

You meet with a senior consultant, Amber Quist, to discuss what kind of publications meet your client's needs. You sit down with Amber behind her desk so you can see her computer screen. Concepts, Inc. keeps a database of information on publications and the costs of different sizes and types of ads for each magazine and newspaper with which anyone in the firm has ever used. A team of college interns keeps in contact with the publishers to make sure the information is up to date.

"What about snowmobiling magazines?" you ask Amber.

"I don't know," she replies. "Do snowmobilers make enough money? I've always imagined them to be more of the motorcycle types, middle income, you know."

"I'm not so sure," you say. "Do you realize that some of those snowmobiles cost several thousand dollars?"

Amber calls up information on her computer. "What size ad were you looking at, anyway? And are you going black and white or full color?"

"We'll have a picture, but we're going black and white to stretch the budget a bit. I'm thinking around six inches wide by three inches tall, or the closest we can come to that with the different column widths. Even though you've been working in the marketing field for a while, you never quite get use to the different ways each publisher measures horizontal space. Some do it by inches, others by the number of "columns," which often vary in width from one publication to another."

"Well, we have *North Country Trails,* which is a magazine with information about snowmobile trails and accommodations in the Great Lakes area," Amber says. She tags the record for printing later. "Hey, what about kayaking or canoeing magazines?" she says. "Agate Point was a stop on the sea kayak race last year."

"Sounds good," you say. "I placed an ad in *Paddle Sports Magazine* last year for a sporting goods client, and they had some response. Tag that one for printing."

"Got it," Amber says. "Any other ideas?"

You think for a moment. "There's a log cabin on Pebble Island. What about log homes magazines?"

Amber searches through the database and comes up with three publications. *Log Cabins, Ltd.* is a magazine that has plans and tips for building log cabins. *Log Home Decor* is an interior design magazine for people with (or thinking about purchasing) log cabins. The last publication listed was *Back Country Living,* a magazine about rustic homes in general, not necessarily log cabins. You ask Amber to tag all three for printing. She sends the command for the computer to generate a report from all tagged records, and you take the following information back to your office.

Title	*Average Reader Circulation*	*Income ($)*	*Column Size (inches)/Type*	*Times*	*Cost ($) Per Insertion*
Back Country Living	200,000	85,000	3 × 3 B&W	1	450
				3	425
				4	385
Log Cabins, Ltd.	60,000	70,000	3 × 5.75 B&W	1	350
				4	310
				8	300
Log Home Decor	50,000	95,000	4 × 4 B&W	1	550
				3	500
				5	485
North Country Trails	280,000	48,000	3 × 3 B&W	1	250
				4	230
				8	200
Paddle Sports	380,000	63,000	3 × 6.25 col. B&W	1	440
				3	420
				5	400

After examining the data on each magazine, you start thinking about some of your other options. You know that Mark Storey wants to run separate ads for each property—one for Pebble Island and one for the Agate Point development. Because these will be different ads, you cannot take advantage of the price breaks for running an ad more than once. The magazines will not allow you to take advantage of the four-times insertion price if you run two ads for the island and two for Agate Point. There's also the matter of timing; you ponder what months would be best for selling property to snowmobilers and what months for kayakers. From what Mark Storey has told you, he receives very few requests to view any waterfront property past October, because most buyers are from further south and do not like to drive in snowy weather.

You sit down at your computer and begin composing the Waterfront Realty document. *So many different choices, and only $20,000 to work with,* you think.

Suggested Next Steps

- Review the guidelines for creating reports as outlined in this chapter.
- Analyze the information about each magazine and discuss how the readers might be attracted to the different properties in your Solution Defense.

- Be sure to include specific information and justifications for your recommendations to Mark Storey within your document.
- Discuss how you arrived at your recommendations in your Solution Defense.

What's the Plan?

In this situation, you are investigating the costs of offering health insurance to your employees. As a business owner, your main concern is the profitability of your company—after all, health insurance won't do your employees much good if you end up having to lay them off during financially lean times. While reading this scenario, keep the following questions in mind:

- What are the basic differences between the different plans described?
- Who are your readers, and what are their motivations and attitudes toward this project?
- What outcomes do you desire from the dissemination of your report?

You are a part owner of Telotech, a communications technology company that has been in business for three years. Telotech is enjoying a great deal of success, and you and the other owners have been able to expand your service area to take in more cities in northeastern Ohio.

With this expansion has come the need for more employees; the company has recently hired new staff and now consists of seven full-time and three part-time positions. Three of the full-time employees are single men; one is a single woman, and the others have spouses and young children.

Now that Telotech is doing well, the other two owners, Jessica Hanson and Ryone Smith, have decided to obtain group health insurance and pay for a portion of the medical and dental coverage for all full-time employees. It fell on you to review different plans and provide a report outlining the benefits and drawbacks of each plan.

You called two insurance agents and had them send you information on small-business group medical and dental plans. The first agent, Gary McLeod, recommended a company called Advanced Health Solutions (AHS) as the best match for your business.

"They cover everything, but with varying co-payments according to whatever medical procedure you use," Gary had said. "It's not like some other plans in which they don't cover anything until you've spent three or sometimes five hundred dollars."

After you receive the information from AHS, you read over the lists of services and co-payments. You jot down what you feel are probably the most common medical services your people will need and how much they will have to pay each time they (and their dependents) use those services.

Medical	Cost	Dental	Cost
Emergency Room	$200	Cleaning and Checkups	$50
Routine Physical	$40	Fillings	$100
Diagnostics (X-rays, etc.)	$80	Oral Surgery	$150
Outpatient Mental Health Service	$65	Orthodontics	60%
Pap Smear	$50	Orthodontic Adjustments	$50
Prenatal Checkup	$50	Dentures	60%
Childbirth	$200	Denture Adjustments	$50

You also do some calculations to figure out how much per month the company will have to pay per employee to buy into the AHS plan.

single male: $65
single female: $72
married couple: $150
each dependent: $50

The other insurance agent, Tom Sanders, recommends an entirely different plan. The company he has worked with most, he tells you, is called Continental Health. Continental works on an entirely different system, according to the materials Tom sent you. The company pays 100 percent of all costs for routine services (e.g., physicals, Pap smears, dental checkups) and all emergency services after the employee has incurred a certain amount of expenses ($400 for a single male or female, $600 for insured families). You see some co-payments, but only for certain expensive services such as magnetic resonance imaging (MRI) and non-emergency operations. The highest co-payment listed was 20 percent of specific diagnostic procedures and non-emergency or elective operations.

The monthly costs for the Continental plan, however, are much higher than for the AHS plan. You calculate the monthly fees.

single male: $95
single female: $92
married couple: $250
each dependent: $70

As you are pondering the many different choices, you hear a knock on your door. You look up to see Christine Centers, an account executive, standing in your doorway. "I wanted to talk to you about the health insurance plans I heard you were considering," Christine said as she took a seat across from your desk.

"We haven't decided on anything yet," you say. "A conversation on what you will or won't receive is premature right now."

"I'm aware of that," she says. "But it was the nature of the plan the company is choosing that has a few of us concerned."

"A few of you?" you ask. *There are not that many employees to begin with,* you think.

"Well, the married people," she says, smiling. "We're kind of outnumbered here, and we are hoping that whatever plan the company chooses, it's not too much more expensive for the families to use."

"Do you have specific concerns?" you ask.

"Yes," Christine says. "For example, Lynn's three kids are both nearing their teens, and it looks like they'll all need orthodontics. Peter and his wife have two children, eight and twelve years old. I have one child who's five, and my husband and I would like at least two more before we're done. So we have people who all make in the mid-thirty- to low forty-thousand-dollar range a year in income with very different needs and obligations than the younger full-time employees. You know, I don't even go skiing anymore, but I know Beth does and she'd want great accident coverage. But she also has more of her income to play with than those of us with families do."

"I see," you say. "Would you rather know how much you'll have to pay each year for normal stuff and things like orthodontics instead of paying as you go?"

"If paying as I go is going to turn out to be more than paying a certain amount, then I'll pay a fixed amount each year," she replies.

"Anything else?" you ask.

"No," Christine smiles and gets up to leave. "I'm really looking forward to having insurance. It's been hard, worrying about whether or not my husband or my kid will get hurt too badly for us to handle it financially."

"I think we'll all be relieved when the plan is in place," you answer.

For a while after Christine has gone, you sit and ponder your situation. No matter what the company decides, there will definitely be very different impacts on and reactions among your employees. You begin to write up what you have come to understand about each plan. All three owners are meeting next week, and you want to convince Jessica and Ryone to make a decision as soon as possible.

Suggested Next Steps

- Review the guidelines for writing reports as outlined in this chapter.

- Consider how much emphasis to give to the benefits and drawbacks of each plan (and the different aspects of each plan). Justify how you present your evaluation of the plans in your Solution Defense.
- In your Solution Defense, discuss how (or if) Christine's input into the health insurance plan issue had an impact on your assessment of the different plans.

The Construction Accident

CONSIDERATIONS

General contractors make sure that a construction job is on time and within budget, and they often find themselves in difficult situations when their employees cause delays and cost overruns. In this scenario, you are faced with a particularly difficult situation that puts your professional and personal priorities in conflict. As you read the scenario, keep the following questions in mind:

- From a professional standpoint, what are the problems and possible consequences of this situation?
- What are your personal feelings about the construction accident, and how should you relate the events of the past few days to your boss?
- What written communication must you create, and what are your options when writing it?

Rock Bay, Michigan, is a small town created during the boom-and-bust logging years. You know it well, having lived in the town as a child. A few years ago, you bought forty acres with a log cabin, and you use this second home as a place to get away on weekends and holidays.

Only a few dozen people live in Rock Bay year round, and most are either retired or own and operate the town's convenience store, laundromat, bed-and-breakfast, and restaurants and bars. Because you live only a half-hour away from Rock Bay, you've come to know the business owners and their families fairly well. You've even dabbled in Rock Bay's politics, providing advice to the zoning council as they struggle to develop the town without losing its rustic charm.

The reason you were asked your opinion about where to allow recreational home development is that the townspeople discovered your occupation—you work as a senior manager for Smith Builders, a well-respected general contracting company. In the course of its thirty-year history, Smith Builders earned a reputation for building high-quality, luxury homes and second homes, what people in the north country call "camps." The term, however, seems hardly appropriate for a Smith-built home. While Rock Bay has seen quite a few glass-and-log structures appear in the last few years, and the locals are well aware of the whirlpool tubs, high-tech kitchens, security systems, and other luxuries inside, they still refer to all houses not occupied 365 days of the year as "camps." The owners of these woods palaces are treated with a polite warmth, as though the dollars spent in the restaurants and grocery store were almost enough to make up for the loss of a good forty acres of woods or lakeshore. Almost.

You, however, enjoy a deeper connection to the year-round Rock Bay residents, given your family's former residence there. Whenever your year-round neighbors see your car in your camp's driveway, it's never more than an hour or two before they appear with loaves of bread, cakes, cookies, or some other treat in the traditional offering to come inside, have some coffee (which you always start the minute you arrive), and trade stories. Today, your neighbor to the east of your cabin arrives, a friendly Finnish woman named Belle Poikkala and her taciturn husband, John.

You greet Belle and John, seat them at your kitchen table, and start the coffee around while Belle cuts and distributes the loaf of cardamom bread she'd brought.

"What brings you up this weekend?" Belle says.

"I have to check up on the Agate Beach road site," you say. "We've been running a little behind on the construction, and Mr. Smith is concerned that we won't be weatherproofed before the snow hits. The clients, the Parsons, are friends of Mr. Smith. Kelley's the foreman," you mention, referring to Terrence Kelley, a Rock Bay native who's one of your best on-site taskmasters. "And he called me and asked me to come up here this weekend."

"Did he say anything else?" John asked quietly. You are a little surprised, because John rarely asks questions.

"No," you answer. "Just that I needed to come up here." You pause, remembering the conversation. "He sounded a little off . . . tired, maybe. Kelley hardly ever asks me to visit a site unless something's either really great or going really wrong, though."

"Really wrong," John says. His large hands, hands that had worked for a logging company and then as a carpenter, looked ridiculous as he delicately stirred sugar into his coffee. You realize that John, who seems so comfortable anywhere that he hardly speaks, is definitely bothered by something.

"We've known you since you were little," Belle starts. She was a volunteer assistant in your kindergarten class, and she'll never let you forget it. "So we're here to ask you for your help."

"What happened?" you ask.

"The construction site . . . ," Belle takes a sip of coffee, "was sort of vandalized."

"What! When?"

"Three days ago," Belle says.

"Three days!" You are immediately confused and upset. "I should have been notified immediately! The police reports have to be obtained, the insurance claims filed, the owners contacted, a new timeline figured out." You stop abruptly when you realize Belle and John are staring down at the table.

"Well, we haven't exactly called the police yet," Belle says. Before you can ask for some kind of explanation, there's a knock at your door. You open it, and Terrence Kelley walks past you, into your kitchen, and sits down.

"You could have waited for me, you know," he says to Belle.

"John started it right off, like I told you he would," Belle retorted. Terrence simply looks at John, who shrugs. You sit down at the table and ask for a full explanation. You find out that the Kasperson twins, two rambunctious thirteen-year-old boys who live near the Agate Beach Road site, decided to play with some of the equipment after the construction crew had quit for the day. They managed to start up a small front-end loader, and accidentally drove it into the side of the house.

"Why were the keys in the Cat?" you ask Terrence, referring to the front-end loader.

"I know, I know," he says, shaking his head. "Company policy is very clear. I should have all the keys to all equipment in my hands before leaving the site. Thing was, I had to run home that day. One of my kids fell out of a tree, and my wife thought she'd broken her leg. Turned out she was just bruised, but I was so shaken up I called Tom and asked him to close down the site at the end of the day." Tom is Terrence's best carpenter; the two have worked together for the last twelve years.

"How bad was the damage?" you ask.

"Well, now that depends on what you do or don't want to know," Terrence says.

"You'd better explain that," you say.

"You know the Kaspersons, right?" Terrence asks. You nod, and he gives you a little time to think about what you do know. The Kasperson family moved to Rock Bay and worked for the first logging company before there was even a town. Once fairly affluent, the family suffered when the woods were played out and the company started selling off land for recreational homes. Mark Kasperson, the father of the twins, works as a part-time carpenter and handy man, while his wife works a few hours each day in the elementary school kitchen. You know that Mark hunts, and sometimes poaches, deer to feed his family. Their situation is not uncommon in small towns like Rock Bay.

Terrence continues. "And you know what Smith's insurance company will do when they find out who caused the damage? File a lawsuit against the Kaspersons. It will put them over the edge, and they've been walking pretty close for a long time. They'll lose their place, where their family's been since I don't-know-when. They'll probably have to move into town and take government assistance."

"We've all been working," John says suddenly. You look up at him and his clear blue eyes lock on to yours. He holds up his hands, and you see fresh cuts and scrapes. "Everyone who can has been at that place, fixing it. In three days, you wouldn't know that it happened."

"Terrence!" you exclaim. "These people aren't insured! If someone gets hurt, we could get one hell of a lawsuit."

"First off," Terrence says. "Not one of these people would ever sue the company. Ever. And you should know that. Second, realize that I didn't have to tell you anything. The Cat could've just rolled into the building, because I didn't check to see if it was secure."

"Why didn't you just say that?" you ask him.

"I don't like lying to you, boss," Terrence says.

"And you think I like lying to Smith?" you reply. "You want me to tell him that the Cat rolled into the side? And give him damage reports that match that conclusion?"

Terrence nods. "It's almost true. The Kasperson kids were just along for the ride. With their feet on both the accelerator and brakes, from what they told me. You know those kids have been there every day after school, hauling scraps and shoveling? They couldn't feel worse about this."

"Let's go to the site," you say. Belle and John get up to leave, making their apologies for bringing you bad news. After they leave, you and Terrence get into his truck and drive out to Agate Beach Road.

The first thing you notice is that the entire construction crew is there, even though they usually have Saturdays off.

"Are you giving them overtime?" you ask Terrence.

"No," he answers. "They're volunteering." You suspect that Terrence is somehow compensating his crew, either with after-hours gatherings at a bar or maybe getting them some extra side work after the building season ends.

You walk around the house, and what happened seems relatively clear. The front-end loader was parked facing the house, about thirty feet uphill from the northeast corner. The kids started the Cat, which rolled down the hill and slammed into the side of the house, near the corner. From what you could see, the impact broke parts of the lower level's wall, some of the nonstructural masonry work that covered the foundation, and cracked the supports where the two walls come together in the corner. The entire house, which was only framed in with its plywood covering on, might have shifted its weight toward the impact area.

To repair the house, Terrence's crew had pulled the plywood off that area of the house, and put in new two-by-fours in the wall. The corner support could not be safely removed without damaging the rest of the house, so the crew had properly reinforced it. You realize the home would easily pass inspection. Everything had been done according to or exceeding the code, and the only real problem was the loss in time and the money for materials. You are relieved to find the house is, for all practical purposes, as sound as it was before the accident.

The crew was working to replace the plywood while continuing the work that should have been completed during the last three days. While half the crew finished the repairs, the other half were busy putting in the subfloor and creating the staircases.

"How far are we behind?" you ask.

"Three days. It looked like six or seven to me before this extra help came," Terrence answers. "We're about two thousand back, because we needed extra lumber, nails, and I had to call the mason back to fix the foundation facing."

"I can't hide two thousand dollars," you say. "I won't."

"I don't expect you to," Terrence answers. "The way I see it, you have a bunch of options right now. You have to give Smith a progress report—I know he's been worried about this place because the client is a potential business partner. You can tell him what really happened, and the police will file the report they're holding on to right now."

"They have a report they haven't filed?" you ask incredulously.

"There's only one police officer in Rock Bay, and he's my wife's cousin," Terrence replies. You nod, no longer surprised by much. "As I was saying, the police report will be there when the insurance company requests it, and it will say that the twins rammed the house, and the insurance company will predictably file suit against the Kaspersons. Your other option is to say the Cat rolled into the house, and we're making repairs that put us back a couple grand and a couple days. I'll take the blame for that happening, because it was my fault the keys were in the thing in the first place. Your other option is to say we've had some cost overruns and delays, and we'll be back on schedule by next week. Take your pick."

You and Terrence say little on the ride back to your cabin. You pick up your briefcase and sit on the couch. Taking out your laptop computer, you open up the word-processing program, and stare at the blinking cursor while you think about your situation. You know that there is no getting around giving Mr. Smith a formal progress report, because he personally asked you to check on the site. You think about the Kasperson twins, and how they could hardly look at you as they worked, almost frantically, at the construction site. You have some tough choices to make.

Suggested Next Steps

- Review the guidelines for making ethically challenging decisions as outlined in chapter three.
- Review the guidelines for creating short reports as outlined in this chapter.
- In your Solution Defense, discuss at least three possible ways you could write your report, the potential outcomes of each of the three actions, and why you chose a specific course of action.

Soil Incineration: Alternative Combustion Methods

Considerations

In this portion of the soil incineration project, you are given the task of researching alternate soil reclamation methods. While this may seem counterproductive to Donnelly Engineering's support for the incinerator, the information is necessary for the final proposal, which must compare your plans for the incinerator with other available technologies. Review the project information from previous chapters, and while reading this installment of the situation, keep the following questions in mind:

- Who are the readers for this project, and what will their understanding of technical information be?
- What are the varied purposes for which your document could be used?
- How do the political aspects of the soil incineration project affect the creation of your document?

A few more weeks, and you will be turning over the completed proposal for the incinerator to the senior management of Donnelley Engineering. Today, you come to work with a bit more enthusiasm for the project than you normally feel. Jack McGregor, your supervisor, has asked you to meet with him about doing some research. You feel good about the meeting and look forward to working on something new.

Jack joins you in the company conference room and asks you a few questions about other projects you are handling. Then he settles back into his chair and says, "We're getting some pressure to do different research on the soil incineration project."

"From where?" you ask.

"I met with a few key people at the township and county levels," Jack says. "I asked them what their major concerns were with this project, and most just don't like the methods we're proposing. It seems that that anti-incinerator group, POL, keeps talking about 'alternative processes' like what we're going to build is some kind of antiquated, depression-era facility."

"What alternative methods are they talking about?" you ask. In the course of your research, you'd occasionally come across references to different ways of removing volatile organic compounds (VOCs) from different types of soil.

"I don't know," Jack says. "That's why I want you to research these other methods, and write up how they compare to our process. I have a sneaking suspicion that these newer methods are either too expensive for our client, or they're so experimental we don't really know how well they'd work. Our project puts the soil through so many different processes, I don't see how one new development can replace this kind of incinerator."

You realize that Jack is, essentially, telling you what conclusions he wants you to reach before you've even started doing the research. You decide not to address the possibility that different methods may, in fact, provide the same results as the proposed incinerator.

"Do you have some information sources in mind?" you ask.

"Not really. Start with the Internet, call companies, I don't know," Jack replies, sounding frustrated. "Research is your thing, isn't it?"

"Yes," you answer. "When do you need this?"

"As soon as possible," he says. "And keep all your original documentation handy, because I bet we'll need to prove how well our facility will work in comparison to any other process when we send out our final proposal."

"Sure thing," you say. "I'll get right on this."

After discussing a few other issues, you return to your office and start up your Internet browser. *Anything I read that's published by a manufacturer is going to be biased,* you think as the computer starts listing sites in response to the key words you've entered. *I wonder how I can find out from some kind of independent source how well these other methods work?*

Suggested Next Steps

- Review the guidelines for creating reports as outlined in this chapter.
- Conduct research into alternative methods for decontaminating soil. Document your sources and research techniques in your Solution Defense.

- Did the pressure to downplay the effectiveness of other soil decontamination processes impact the way you gathered or presented your information? Discuss your observations of any impact in your Solution Defense.
- Consider the reliability of information you pulled from the Internet and other sources. Document and discuss any potential bias in the information as part of your Solution Defense.

CHAPTER 9

Proposal Writing

In the past eight chapters you have learned how to write everything from a standard, internal memorandum to an external business report. In this chapter, you will learn the rudiments of proposal writing, a relatively common type of technical communication in the workplace. Although many of the forms and styles covered thus far in the book use elements of persuasion, proposal writing is very closely tied to persuasion. When you write a proposal—internal or external—you are attempting to persuade a group of people that your ideas, suggestions for change, or recommendations are not only sound but are also potentially better (for the company or the client) than any others. Thus, in proposal writing you face not only the challenge of developing good ideas and writing them clearly but the additional challenge of "selling" your ideas to an audience.

Proposals are responses to (1) a request for services or (2) your perception of a need. Indeed, virtually all types of proposals fall into one of those categories. Before evaluating both categories of proposals, consider some of these points.

1. Because you are attempting to persuade an audience in this document, always spend the time necessary to learn as much as possible about them. For internal documents, this will often be less difficult. For external documents, you need to research clients to understand as much as possible about their needs.
2. Because the proposal is a persuasive document, rely upon your expertise (or the expertise of your employees) to "sell" your plan. Avoid references to your competition (e.g., D & B Construction, Inc. can always beat the estimate and timetable of Rundell Constructors) because you are in the business of promoting your idea.

3. Regardless of the type of proposal (internal or external) use graphics to depict complex information (see Guide to Creating and Working with Visuals) and the following style guidelines:
    - Use headings to move through the various categories of information.
    - Use white space around the text to increase readability.
    - Use bulleted or numbered lists to break up items in a sequence or long lists of information.
4. Develop an outline for the proposal before you begin writing. Failure to do so can result in the omission of important information. Study the problem thoroughly and plan your response before you begin writing.
5. Finally, be reasonable in your response. If you are responding to something as complex as a government request for services, or if you simply perceive a need for change in your office, do not over- or underestimate your abilities. Do not suggest that a job can be completed in three months when, in all likelihood, it will take six months. Do not indicate that you can complete a job for $5,000 when you know costs may run considerably higher. Particularly when you are responding to external requests for services, proposals can be binding legally. Be honest and reasonable in your assessment of time and money and, most importantly, in your ability to do the job.

PROPOSALS BASED ON REQUEST FOR SERVICES

In the workplace today, perhaps the most standard mechanism for triggering a proposal is the external request for services. Often a business—be it a private business, a large corporation, or even the federal government—seeks the unique expertise of others to complete a required job. Companies do not always have employees "in-house" who can perform specific tasks. For example, imagine that you and a partner opened a small business specializing in software development, and within a year your sales figures were so high that the two of you decided to open a branch office in a small city approximately 100 miles away. You do not find acceptable office space in the targeted area, so you publish in the local and surrounding newspapers a *Request for Proposals.*

The request for proposals, or RFP, is a general call published by companies, individuals, and large corporations seeking expert services. In the sample case above, your published RFP would invite construction companies to "bid" on the job by presenting you with a list of credentials for employees, past successes in building and construction, a timetable for completion of the project, and, of course, a budget. The RFP certainly is not limited to a small scale construction job. The government regularly publishes RFPs for specialized services involving robotics, computer software installation, large-scale construction and development projects, and so on. Consider the following sample RFP, the sort often published by state departments seeking specialized services:

Request for Professional Technical Services Iowa Department of Transportation Technology Division

The Iowa Department of Transportation (IDOT) requests proposals from companies or individuals interested in providing the following services. IDOT seeks proposals, specifically, from companies or individuals who can provide qualified consulting services. IDOT anticipates that a maximum of twelve consultants (individuals or within consulting firms) will be placed under contract for the services outlined below.

Claim Evaluation

IDOT seeks qualified consultants to analyze contractor claims for additional compensation and compare those claims to specific contractual guidelines. After reviewing contractor claims, the consultant will write a report outlining findings and submit that report to IDOT. The consultant will also evaluate the validity of claims and advise IDOT accordingly.

Please note the following materials to be submitted with your response to this request for services:

1. Include your resume or vitae curriculum detailing your knowledge of the field, your experience, and your general capability to evaluate contractor claims and compare them to specific contractual guidelines.
2. Include in your response your general understanding of the problem, how you would evaluate claims and respond to them, and how much time, in general, a standard evaluation might take. Include, as well, your fees for services and any other expenses that you may incur.
3. To be considered for this job, you must have familiarity with IDOT's specifications and practices and be able to demonstrate this familiarity. In addition, you must have specific background experience in either mechanical or construction engineering.

Interested consultants should submit full proposals no later than 5:00 p.m., Friday, June 20, 1998.

Evaluate this sample RFP for a moment. Notice that the company (in this case a state department of transportation) seeking qualified individuals specifies very clearly not only what type of background the respondent must possess but also a general outline of information that must be included with the proposal. Guidelines for writing the proposal follow in this chapter, but in general remember the following important point.

TIP

> If the RFP gives the respondent specific guidelines for writing the proposal, follow those guidelines completely. Do not create your own arrangement of specific parts of the proposal if the company, corporation, or department very specifically outlines the format to be followed.

In addition to proposals that stem from the RFP, sometime proposals are generated by the individual in response to a specific problem.

PROPOSALS BASED ON A PERCEIVED NEED

Sometimes individuals perceive a need, either within an institution (e.g., company, small business, or corporation) or within society in general. If you have a good idea for change and believe you have the expertise and capability to initiate that change, you might write an unsolicited proposal based on a perceived need. For example, suppose that you work as a technical communicator in a large company that develops software programs. You, as the lead technical communicator, supervise a small staff of assistants who help with graphic design, layout and formatting, and assorted other tasks connected with the creation of manuals to accompany new software packages. Frequently you or members of your staff are sent out "on location" to observe users in the field and perform usability tests on clients who read the manual and attempt to use the software. Routinely you must explain to your staff members how to fill out and submit travel documentation for reimbursement. The documentation is not all that simple. Employees must use certain codes in state and other codes for out-of-state trips. All receipts must be submitted with appropriate paperwork; mileage and mileage rates, which vary according to the trip must be recorded and so on.

One day, after you've explained the procedure for what seems like the hundredth time, you sit down and write an internal, unsolicited proposal to your supervisor, the head of the communications department. You believe that the company needs a guide—a handbook—for travel documentation and reimbursement. You believe that you can write this guide efficiently because you not only know the policies well, but, of course, because you are a technical communicator.

This sample situation demonstrates how, sometimes, proposals are not related to specific requests for services. Often individuals perceive needs and respond to those needs by offering services. Regardless of the *reason* for writing a proposal, however, you must consider carefully where that proposal is going, either internally to a colleague or supervisor or externally to a client.

Internal Proposals

As you learned earlier in this book, internal documentation usually takes a memo format. The same is true of the internal proposal. While still a persuasive document, the internal proposal is usually addressed to someone who already knows your work (as you saw in the previous scenario), so you do not need to include a resume or curriculum vitae with the memo-style proposal. In fact, internal proposals are most often written in response to a perceived need within an organization, so they are typically shorter and often involve less complexity than the external proposal. With only a few exceptions, you do, however, need to follow the general format for proposals. Refer to upcoming Guidelines for Writing the Proposal below.

External Proposals

External proposals, those written in response to requests for services, are longer documents that follow essentially the same format as the long report discussed in Chapter 8. Because these proposals *leave the home company or organization,* they tend to be more complex and include more parts than the internal (or memo-style) proposal. Refer to the following Guidelines for Writing the Proposal for specific information on how to develop and format both the internal and external proposal.

GUIDELINES FOR WRITING THE PROPOSAL

In general, both internal and external proposals contain similar parts. The following list contains the components of the proposal. Note that some parts are excluded from the internal proposal.

- title page (exclude from internal proposal)
- letter of transmittal (exclude from internal proposal)
- table of contents (exclude from internal proposal)
- introduction
- statement of problem
- proposed solution
- qualifications and expertise
- budget and timetable
- appendix (optional in both internal and external proposals)

Title Page

The title page, which immediately follows the cover of your external proposal, contains your name, the date, the title of your proposal and the person for whom the proposal is intended. Refer to the sample external proposal in Figure 9.1 on page 257 for an example of the title page.

Letter of Transmittal

The letter of transmittal is particularly important in the external proposal because it is written directly to the person (or persons) who requested your services or expertise. Because internal proposals are written in a memo format (and thus addressed to a specific party), the letter of transmittal is redundant in the internal proposal. For an external proposal, write the letter of transmittal according to the standard conventions of letter formatting covered in Chapter 2. The letter, which directly follows the title page, previews the proposal and suggests possible recommendations to follow. Refer to the sample external proposal on page 258 for a look at the letter of transmittal.

Table of Contents

While the table of contents is typically excluded from shorter documents such as the internal proposal, all external proposals need a table of contents, a valuable tool for readers who seek quick information and a glance at the highlights of your work. Place the table of contents directly after the letter of transmittal, and arrange the material according to the major headings you have used in the proposal body (the *Statement of Problem* and the *Proposed Solution*). Use the exact wording in the table of contents so that readers will readily make connections from the table of contents to the body of the proposal. In addition, follow these general guidelines:

- *Parallelism.* Make both the headings in the body of your proposal and the headings in your table of contents parallel. For example, if you begin using headings that follow an "ing" pattern (*Evaluating Contract Specifications* or *Determining a Timetable for Completion*), do not switch to a different heading style (*Contract Specification Evaluation* or *When the Job will be Completed*). Changing heading styles can be very confusing for the reader.
- *Page Numbers.* Use consecutive page numbering in the external proposal and reflect those page numbers in the table of contents.
- *Boldface Type.* Use boldface type, underlining, or italicized type to indicate major breaks in the proposal body and in the table of contents. Use adequate white space so that the table of contents is easy to read.
- *Attachments.* At the end of the table of contents, indicate all attachments or appendices. Give all attachments either a page number or an appendix letter (e.g., Appendix A, Appendix B).

Refer to the sample external proposal on page 259 for a sample table of contents.

Introduction

In both the external and internal proposal, begin your document with the introduction, an overview of your proposal that emphasizes your understanding of the problem, your brief recommendations, and the primary points you will cover in the body portion of the proposal. The introduction, as such, is relatively brief. In the external proposal, the introduction should not exceed one page; in the internal proposal, the introduction should not exceed two paragraphs. Refer to the sample proposal on page 260 for an example of the introduction.

Statement of Problem

With the statement of problem, contained in both the external and internal proposal, you begin the body portion of the proposal. Though obviously different in length and detail, this section provides the reader with (a) your understanding of the problem and (b) specific details that support your sense of the problem. This section, then, contains both major headings (main points) and subheadings (minor points). Headings are reflective of the primary information and are parallel in structure. Whether you are responding to a request for services or a perceived need, make certain that you explain to readers your understanding of the problem. Detail specific aspects of the problem as a kind of "set-up" for the proposed solution section that will follow.

Proposed Solution

Immediately follow the statement of problem section with the proposed solution, the section (within the body portion of your proposal) in which you specify how you believe the problem can be solved, what steps must be taken to achieve a satisfactory solution, and who will be effected by the proposed solution. Use headings and subheadings that reflect the detailed breakdown of the problem you used in the statement of problem section of the proposal.

Qualifications and Expertise

In the proposed solution section, you suggested specific steps for "fixing" a problem. Whether the audience for the document is external or internal, you must explain why you are particularly qualified to undertake the recommended solution to the problem. In this section, therefore, discuss your specific qualifications for the job, your past experience in handling similar problems, and, if relevant, the expertise and qualifications of the employees who will assist you. If you are responding to an RFP, include resumes for the personnel who would likely participate in the job if the proposal is accepted. This is really your opportunity to "sell" yourself and your company. Follow these guidelines as you write this section:

- *Be honest.* If the RFP specifies someone with five years of experience in a given field and you have only three years, indicate that you have less than the specified number of years of experience. You can, however, indicate how additional training, education, or other experience might fill that gap.
- *Be selective.* Include information that is relevant to the job request. If you are writing a proposal in response to a need for a groundwater contamination specialist, indicate education, expertise, and experiences that are relevant to that job. If you worked as a journalist before you earned your degree in chemistry, you need not list that information.
- *Be certain you can do the job.* Respond to a request for services if you are sure you (or your employees) have the expertise necessary to com-

plete the job. If you operate a small construction company, you may have difficulty undertaking a major construction project for a nationally known chain of retail stores. Be realistic before you respond to requests for services.

Budget and Timetable

If being honest and realistic are important points to consider under qualifications and expertise, they are equally important points to consider under this heading. Particularly when you are responding to requests for services, be very realistic about (a) what a project will cost to complete and (b) how long it will take to complete. In fact, most sensible and successful proposal writers often over-estimate on the timetable. If you believe, for example, that you and your staff can complete a project in six months, think about whether or not that would be true if key members of the team became ill, if materials were not available for some reason, or if a death in the family pulled an important member of the team away from the project. Add a few weeks to a month to timetables so that you not only have a cushion should problems arise, but also, more importantly, so that you can impress your employer when you complete the project early.

The same holds true for the budget. If you believe that you will need $15,000 to purchase the materials to build a municipal playground in the center of town, consider some of the hidden costs. Once you respond to the request for services with a specific figure, you are committed to that number. Be realistic as you establish a budget for proposed projects.

TIP

Budgetary items need not include only what something costs. You might also note potential savings for the company. For example, a cable modem provides Internet access without using a phone line. Your company could save the monthly charges associated with a separate phone line for Internet use.

All of the aforementioned sections (the statement of problem, the proposed solution, qualifications and expertise, and budget and timetable) all go within the body portion of your proposal. To make these sections as successful and readable as possible, follow these guidelines:

- Move from general ideas to specific, detailed information as you draft the paragraphs.

- Use bulleted or numbered lists to present important information (as in the proposed solution section).
- Use adequate white space in the body portion of the proposal. Too much text density can cause readers to skim over important information.

Refer to the sample external proposal on page 260 for a student-written example of the body portion of the overall proposal.

Conclusion

The conclusion of the proposal—whether it is internal or external—provides readers your overall recommendations and suggestions for change. The conclusion of the external proposal is typically longer than the conclusion of the internal proposal, though both summarize, in particular, the recommendations for solving the problem. Include in the conclusion a numbered list of recommendations so that readers are reminded, one last time, of your suggestions. Refer to the sample external proposal on page 262 for an example of the external proposal conclusion.

Optional Features: Attachments and Appendices

In some external proposals, you may need either additional attachments or appendices. Because both attachments and appendices are optional, follow these general guidelines when deciding whether or not to include this material. Place attachments or appendices at the end of the proposal, immediately following the proposal conclusion. An appendix, though important, is typically something that is either not appropriate or too lengthy or detailed to include in the proposal text. For example, in an external proposal in response to a request for services, your resume and the resumes of your employees should be included as attachments.

Appendices are typically used by readers seeking specific or specialized information that might not be relevant to a more general readership. As a general rule, restrict appendices to no more than six additional pages of material. (Note: A group of resumes, as attachments, may exceed this number.) A proposal with too many appendices may seem overwhelming to many readers.

Regardless of the proposal type (request for services or perceived need) or style (external or internal), remember always to consider the needs of your audience, to write coherently and succinctly, and to use the active voice. The classroom exercises following the sample paper on page 262 will give you practice writing internal proposals and thinking about the components of the external proposal.

Title page

PROPOSAL TO IMPROVE VENTILATION
IN NORTHERN UNIVERSITY'S
UNIVERSITY CENTER SMOKING ROOM

Respectfully Submitted to the Dr. Scott King,
Dean of Students
Northern University

December 1, 1999

By

Margary Ross, Chair
University Center Renovations Subcommittee

FIGURE 9.1 Sample Proposal

Letter of transmittal

December 1, 1999

Dr. Scott King
Dean of Students
Northern University
Montcalm, NE 56663

Dear Dean King:

In accordance with policy 531C, the Campus Renovations Committee will address annual renovations on January 14. Our subcommittee on University Center renovations proposes improvements in the University Center's designated smoking room.

Considering the university's limited funds and the dozens of submitted proposals, we have developed a cost-effective plan for improved ventilation with benefit for smokers and non-smokers alike. Please read the following pages before allocating renovation funds for next year.

Respectfully,

Margary Ross

Margary Ross
Chair, University Center Renovations Subcommittee

FIGURE 9.1 Sample Proposal *continued*

Table of contents

TABLE OF CONTENTS

FIGURE 9.1 Sample Proposal *continued*

Introduction

Introduction

In 1992, Northern University renovated its University Center cafeteria and lounges, creating separate smoking and non-smoking areas. The room designated for smokers is separated from the rest of the area by a glass-and-steel wall. The room contains several tables and two couches and is located on an outside wall with several windows.

Within weeks of the first semester after these renovations, it became obvious that the new ventilation system was grossly inadequate. Student workers complained about the air quality in this room, which they are required to clean regularly.

This proposal examines the need for improved ventilation in the University Center smoking room. It also balances cost and benefit issues and details several options for air quality improvement.

Statement of problem

Statement of Problem

With dozens of students using the University Center smoking room daily, the air pollution problem is affecting smokers and nonsmokers alike. The smoking area takes advantage of windows and existing air exchange ducts for ventilation. The ducts, however, are not sufficient for the number of smoking room users. A three-foot thick cloud of smoke hovers near the ceiling—further evidence of the ventilation system's inadequacy.

Visible smoke affects the health of smokers, passers by, student workers, and cafeteria employees. The simple opening and closing of the door allows smoke seepage into nearby nonsmoking areas. Inside, hazardous carcinogens are obviously lingering. This is visually unappealing, as the major traffic corridor between the cafeteria and adjoining buildings runs directly by this smoking room.

Cafeteria employees have made a formal complaint to the Food Service Department, demanding better ventilation for this room. The complaint indicates that student workers who are nonsmokers often refuse to work in the room as part of their regular cleaning duties.

1

FIGURE 9.1 Sample Proposal *continued*

Proposed options

Proposed Options

Renovation of Existing Air Exchange System

The capacity of the air exchange system in this room needs to be doubled. This involves tearing down the existing drop ceiling, installing new duct works, and adding another fan to blow air out of the building (which will require another opening in the building's exterior wall). According to Bellows Heating and Cooling, the contractor hired to complete the original system, these renovations will cost in excess of $7,000.

Smokeless Ashtrays

Available at local stores, smokeless ashtrays operate by battery to filter air through its lid while the cigarette rests in the tray. When closed, the ashtray targets ash odor. Each tray costs $15.96 with replacement filters available for $5 each. The smoking room would require six to ten smokeless ashtrays.

Box Fans

A ten-inch, three-speed, standard fan costs $8.38 at a local hardware store. If placed so that it blows air out the smoking room windows, this fan would speed removal of cigarette smoke from the room.

Air Filters

Local vendors sell traditional air filters ranging in price from $68 to $160. These filters vary in the amount of smoke, odor, and allergens that they remove from the room. Additional costs include a filter change every two to three months. Because of the volume of smoke in the room, our committee recommends monthly filter changes. Filters, available in packs of two, cost from $5.96 to $11.95. These are the only products in our proposal with measurable results. The manufacturers claim the filters process 99 percent of the smoke, dust, and dirt particles.

2

FIGURE 9.1 Sample Proposal *continued*

Qualifications and expertise

Qualifications and Expertise

Members of the University Center Renovations subcommittee are employees at the University Center and two students who also work at the center.

Members interviewed various University Center employees and conducted interviews with students using the smoking room. Members researched available smoke removal devices and their prices by obtaining estimates and visiting stores selling such devices.

Budget and timetable

Budget and Timetable

We recommend immediately installing the following devices as a stop-gap measure until more permanent renovations are made.

Ten smokeless ashtrays @ 15.96 each	$159.60
Replacement filters (20 @ $5 each)	$100.00
One box fan @ 8.38	$8.38
Two air filters @ $160 ea.	$320.00
Replacement filters (6 packs @ 11.95 each)	$71.70
TOTAL	$659.68

Conclusion

Conclusion

These measures should decrease the smoke stagnation problem until the end of the semester, when the university should double the air exchange system's capacity at a quoted cost of $7,000.

3

FIGURE 9.1 Sample Proposal *continued*

Exercises for Classroom Discussion

1. You work for a midsized public relations firm with a staff of about sixty people. You and several of your female co-workers have for some time discussed, during a coffee break, just how nice it would be to work on a compressed time schedule so that you could have an occasional long weekend at home with your children. Under the plan you've discussed, employees would be permitted to add one hour of work to each day (beginning at 7:00 a.m. rather than 8:00 a.m.) for two weeks, so that at the end of the second week, they would have a Friday off. The four of you decide to write an internal proposal to your supervisor proposing the new system.

2. You are a member of the student government at your university. Increasingly, you and your friends notice the number of students who complain about the rising tuition at this institution. While you and the other members of student government cannot control the tuition, you can propose a change to the current tuition structure. You and a subcommittee of three other students decide to research the flat-rate tuition plan used at some colleges and universities and propose to the administration the viability of the flat-rate tuition plan at your institution.

3. You are a relatively young technical communicator at a midsized public relations firm specializing in advertising for industry. Your responsibilities include writing, graphic design, and desktop publishing of company manuals. You are probably the youngest member of the staff at the firm, and you are becoming increasingly frustrated by the number of paper memos that circulate throughout the company at any given time. While a few members of the staff use e-mail, most still prefer the paper memo. You meet with your supervisor to discuss a plan you have for offering training sessions on e-mail and other computer applications to those employees that have e-mail service. Your supervisor is intrigued by your willingness to provide the workshops and asks you to write an internal proposal detailing how you will make e-mail instruction easy to understand for even the most resistant on the staff.

4. Over lunch, you and a small group of friends compare notes on some of the faculty you have had at your university. You complain about some, praise some, and speculate how great it would have been to have some of the information you all shared before you took classes. One member of the group, who is a member of a student organization on campus that works with fraternities and sororities seeking meaningful campus-related projects for the academic year, suggests a proposal writing campaign to the university administration. Why not, the student asks, have a group of people devise a survey of faculty that the administration could approve? The survey could be handed out in every classroom during the last week of the semester, and the results could be tabulated and published on the Internet prior to the next semester. You all nod in agreement and begin brainstorming the proposal that will have to be accepted by the administration.

5. You are a member of a consulting firm that plans to respond to a request for services. Specifically, a local university, with a population of approximately 9,000 full-time students (about half of whom are commuters) wants a study on the feasibility of building a multilevel parking structure in the heart of the campus. Apparently increasing complaints of lack of parking space have led the administration to seek professional advice on the potential for such a parking structure. You must write an external proposal for this project, and the university offers few guidelines on drafting such an instrument. You and your group, therefore, decide to tackle the preliminary document by (a) assessing the qualifications of the group, (b) brainstorming the problem, (c) discussing the potential size and structural limitations of the parking garage, and (d) the expressing concerns in the RFP that the building blend into the campus proper.

SCENARIOS FOR TECHNICAL PROPOSALS

Marni Proposal

CONSIDERATIONS

Many times, technical writers have a great deal of difficulty convincing their colleagues that clear, concise language is more effective than attempting to impress a reader with industry-specific jargon. In this situation, you have all the information you need to write a grant proposal; however, you're not quite sure you even understand all of it. As you read this scenario, keep the following questions in mind:

- For what purpose does your company want grant money?
- What is the process you must go through to have this grant accepted?
- What are the relationships between different people and departments in this company, and how do these relationships affect the grant proposal?

You have recently been hired as a writer for Economic Developers, Inc. (EDI), an area firm that lends money to businesses and is a Small Business Center (SBC). SBCs offer advice and important services to small business owners and help people get through the legal and regulatory tangles involved in starting a company. You are part of a team that writes proposals, reports, and handles other communication.

You have a morning meeting with Bob Tassin, manager of EDI's loan department. You sit down in a conference room together, and Bob begins,

"We have a grant to get out in the next few days, and I'm supposed to fill you in on it."

You feel your stomach tighten a little bit—this is the first project you've been asked to head up—because all you've done so far is some support writing for a report. You grab your notepad and await more information. "It's a grant for loan funds," he says. "We're asking for about $800,000 to recapitalize our loan fund."

"Hold on a second," you say, scribbling furiously. "Please fill me in on the loan fund. I've only heard sketchy details up to now."

"Well," Bob answered, "Mary LaPorte and I handle the lending end. Mary has about thirteen years of commercial lending experience, and I have about seventeen all told. We started the loan fund here about four years ago and have loaned out about two million with only one default. But we got the money back when the company went through bankruptcy." Bob pauses while you catch up on your notes.

"The biggest loan we ever made was $250,000 to a business for new machinery. The smallest one was $500 for one woman to travel to a show where she hooked up with a national vendor. I think she created patterns for craft stuff. You know, stuffed animals and potpourri things."

You smile and nod. "For what other things do you loan money?" you ask.

"Oh," Bob thinks for a minute, squinting behind his steel-framed glasses. "We've lent so businesses can buy a new building, remodel, or add on to the one they have. We've also given out loans for new inventory—sometimes people change their product lines and can't afford all the new materials. Once, we lent to several manufacturers so they could pay the membership fees to join a network for production and shipping. Oh, and then there's employee training programs, money to start new businesses, of course, and consulting for things like new product design or product packaging. That's about all we've ever done, I think."

Your hand begins to hurt a bit, but you don't have time to flex your fingers because Bob is still talking. "We only have about $110,000 left, which will stretch pretty far if we can get some area banks to cofinance with us. We've done that eleven times so far, working with three different banks. Of course, they get their money back first if the business defaults. That's only happened once . . . "

"You mentioned it, yes," you say, not noticing Bob's frown at your correction. "Recouped from bankruptcy."

"Very good," Bob says with just a hint of sarcasm. You make a mental note not to correct him again. "What else do you need to know?" he asks.

You try very hard not to sound rude, because Bob hasn't even given you the guidelines for the grant yet. "Well, what does the grant application require?" you ask.

"Oh, yeah," Bob says, handing you the manila folder he's been rolling up in his hands for the past ten minutes. "Here it is."

You open the folder and scan the proposal guidelines.

Small Business Development Grants

Marni Foundation for Rural Economic Development Proposal Guidelines

PURPOSE: The Marni Foundation announces its fifth annual round of grants that assist rural economic developers in their efforts to support the creation and survival of small businesses. Grants, ranging from $10,000 to $1 million each, may be used for small business counseling, technical assistance, revolving loan funds, and related administrative costs such as staffing, advertising of services, and so on. Consideration will be given to complete, detailed proposals focusing on small business development, especially those that propose improving the technology used by businesses in rural areas.

PROPOSAL: All proposals must include the following information:

- institutional capacity
- region to be served
- detailed explanation of project to be funded
- timeline for project
- budget

On the cover sheet include the company's name, address, and phone numbers and a BRIEF (not more than 150 words) summary of the proposal.

Proposals are not to exceed four typed pages (12-point type minimum) excluding the cover sheet. Proposals exceeding the four-page maximum will not be considered.

"What do they mean by 'institutional capacity'?" you ask.

"Jargon for the company's history, amount of staff, services, and so on. They just want to be sure we're not some fly-by-night operation," Bob answers. "There's a file on the network computer, labeled 'EDI Description'," he says. "We give a printout of that file to people, like journalists or prospective clients. It's pretty good." You remember that one of your "back burner" projects is revising that not-so-good document.

"How about 'region to be served'?" you ask. "Is there another file on this?"

"No," Bob answers, "but there should be. Ask Laura." You know Laura Peters—the company's researcher and statistician. She also manages the computer systems.

"That should be enough to start," Bob says rising from the chair. "I'll be in meetings today to hash out the budget and timeline, so I'll get back to you tomorrow on those two sections."

"One more thing," you say. "Is the Marni Foundation heavily into economic development?"

"Why do you ask?" Bob says.

"I'm just trying to figure out how much technical language to use," you answer.

"I'd keep it really simple," Bob answers, then slightly closes the door while leaning towards you. "Just between you and me," he says in a low voice, "the last proposal didn't make it past the first readers. They didn't understand it."

"Thanks for the information," you say. Most of the company's documents that you've seen have been full of technojargon and extremely long sentences, so you are not surprised. You now have time to go over the company description and try to make sense of it. Turning to your computer, you find the "EDI Description" file, which reads:

ECONOMIC DEVELOPERS, INCORPORATED

• Business Counseling • Loans • Technical Assistance

EDI is a combination firm—our Small Business Center and Loan Programs are nonprofit, while our Technical Assistance services are provided as fee-for-service, although fees are often subsidized.

Located in rural West Virginia, the EDI strives to assist the economic transfer from resource-based (logging and mining) industries to a diverse manufacturing and service industrial foundation.

continued

EDI has housed the Small Business Center for over four years and has four qualified business counselors with a combined business counseling experience of over 46 years. As of the first of the year, the SBC has assisted twenty-two businesses, helping to create an additional fourteen firms. Services the SBC offers are basic accounting training; computer consulting and training; assistance in meeting local, state, and federal regulations for small businesses (including taxes, social security, and insurance); referrals to legal services; in-house employee training on total quality management and workgroup skills; advertising advice and assistance; and market research and assessment.

The Loan Programs offer financing from $500 to $250,000 to qualified applicants for a variety of uses, such as inventory purchasing, company buildings, new technology, and employee training.

EDI offers technical assistance to manufacturers in need of cutting-edge technological advances and alternative production measures to remain competitive. We help manufacturers, those not in direct competition with each other, to pool resources to reduce shipping costs and share new technologies and techniques. The Technical Assistance Program has created a manufacturers network, in which the managers and floor supervisors of related companies come together to increase their competitiveness. As of the first of the year, Technical Assistance has facilitated the formation of networks in the secondary wood products, maple syrup, and metals industries with a combined membership of over twenty-three firms.

After reading the description and comprehending only part of it, you sigh and begin making notes, trying to translate this economic development jargon into more understandable terms.

Next, you ask Laura for information on the region. To your surprise, she sends you the following information via the company's computerized mail:

Hope this helps! —Laura

WV's major industries: resource based (logging and mining). The last coal mine in our immediate area closed in the 1960s, leaving much of the northwestern part of the state without a major industrial base.

There are 55 percent fewer manufacturers in WV than in other states.

Unemployment ranges from 10 percent to 23 percent during the tourist season (May through August) and rises to 12 percent to 30 percent after the tourist season ends. (Percentages are per county.)

There are 105 major banking institutions, which lent an average of 1.2 million dollars each last year.

Small businesses are growing at a rate of 3.4 percent annually. Mostly service industries (e.g., tax preparation, accounting, legal services, consulting, and architectural firms). Many unique ventures such as an artesian spring water bottling company, a rustic furniture manufacturer, deer blind manufacturing, and sophisticated cardboard packaging producers.

100 percent of EDI's clients receiving loans were denied financing from a commercial bank.

Over 350 people attended a recent seminar on starting a small business.

The next day, Bob sends you e-mail on the budget and timeline for the project.

Here's the information. If you have any questions, e-mail me back. —Bob

continued

Budget:	$800,000 for loan money
	$30,000 for staff costs (partial salary for Mary and myself)
	$20,000 for technical assistance subsidies (we're encouraging all borrowers to get some counseling and other help)
	$2,000 for advertising and printing applications
Total:	$852,000

Loans will be from $20,000 to $150,000 for start-up and new technology purchases only. We will charge 8 percent, unless cofinancing with another bank (we hope to do this at least five times with these new loan funds). The partner bank may want a different rate.

Timeline: The loans will be reviewed and approved in rounds. A general timeline is as follows:

From Date Loan Money Received	Activity
30 days	Advertising campaign begun
60 days	First round of applications by staff
90 days	Lending board meets—makes final approval or denial
120 days	First loan checks issued to borrowers
160 days	Evaluation of borrower purchases and timely payments
180 days	Second round begins—process repeated until all money is loaned out.

It's been a few days since you're initial meetings on the Marni Proposal project, and you are beginning to feel more confident about your draft. Even though it's five o'clock, you continue to type diligently—the final draft of the proposal is due soon and you have not been able to find enough quiet time to really work hard on it until now. You make a quick phone call to the take-out restaurant next door and place a delivery order, knowing that the food is so greasy you will probably be able to hear your arteries slam shut, but you really need to eat soon. Because you broke your concentration anyway, you decide to check your e-mail before going back to the proposal.

Your heart skips a beat when you see you have mail from Rebecca Sims, the vice president for operations. Her message reads:

> I know this is rather late in the process, and I haven't been able to make the writers' meetings lately, but I am concerned about one aspect of the Marni Foundation proposal.
>
> Marni doesn't always fund staffing costs, and I would like you to include a justification for the money requested for 25 percent of Bob's and Mary's salaries. You might include language to the effect that almost all of their time will be devoted to administering the loan funds, with the remainder of their salaries being paid by the company's general operating funds.
>
> I will be present at the review of the proposal if you have any additional questions on this matter. Thanks for attending to this on such short notice.

You rethink some of your decisions on this proposal, contemplating how to accommodate all the different requirements placed on you by not only the foundation but also your own company.

Suggested Next Steps

- Review the guidelines for writing proposals as outlined in this chapter.
- In your Solution Defense, document all your decisions regarding the translation of technical information into more simplified terms.
- Discuss your decisions concerning organization and emphasis in your Solution Defense.

File Access from Anywhere

Considerations

Your entire job is solving problems, and in this situation you are trying to find a way to provide a valuable resource to your employees. As you read this scenario, keep the following questions in mind:

- What is an FTP site and why does your company need one?
- What are all the steps you must take to establish an FTP site?
- Whose cooperation do you need to complete this project?

As an assistant manager for Solutions, a well-established computer technology company, your main responsibilities are coordinating different technical staff on various projects. During the last year, you have managed the combined talents of software programmers and graphic artists to produce multimedia CD-ROMs. Recently, you were promoted from the multimedia projects area to the network projects division.

Networking was a new area to you. You were familiar with local area networks (LANs), in which several computers, usually within a building, are connected to a central file server. This allows users to share files, send interoffice e-mail without having to access the Internet, and share resources, such as printers and scanners that might be across the room, down the hall, or even on a different floor than the user's workstation.

Your new projects, however, deal with wide area networks (WANs), which involve connecting two or more LANs in a variety of ways. You are in the middle of a WAN project in which you coordinated several technician teams to first build individual LANs, then connect each LAN using routers (equipment that knows where each LAN is located on the network and sends data to the appropriate LAN server) and what you refer to as "pipes," which are fiber optic or regular telephone lines leased from various phone companies.

Today, you have a project meeting, in which the team leaders from each technical group come together to discuss their progress. Your job is to track each team's activities to determine when the project will be completed, to handle any issues that involve contacting the client, and to gather company resources to ensure each team has all the equipment and assistance they need.

You arrive at the meeting room to find the team leaders already assembled and helping themselves to the coffee and donuts on the table. Because you work with many different teams, you quickly run through their names in your head before they notice your presence: Don Glads, head of the cabling group responsible for the physical computer-to-computer connections; Jeannie Wraith, who handles the installation of the networking equipment; Rex Beck, who is in charge of all the configuration issues involved in connecting multiple PCs to the server, printers, and other shared devices; and Sean Hayworth, the WAN engineer who oversees the connection of each LAN to the entire network.

The meeting progresses smoothly. Don assures you that all the necessary wiring in each of the three locations has tested perfectly. Jeannie and Rex have completed about twenty percent of the installation and configuration of all the necessary equipment. Sean has set up the central WAN server and is awaiting completion of the individual LANs.

"I have run into one particular problem at these sites," Jeannie mentions.

"What?" you ask.

"Well, the computers in each area are widely different from those in other areas," Jeannie says. "It seems that when we did our site surveys, we didn't go into each office and check out what computer was in there."

"This means that when we put in the NIC, we can't always use the same kind and we have to try different drivers," Rex explains.

You know that a NIC is a network interface card, a component the installers put into the computer and plug into the wiring. A driver is a small computer program that establishes communication between the computer and the NIC. There are hundreds of drivers available for computers with different configurations—without exactly the right driver, the workstation cannot communicate with the network.

"So we're looking at more time for installation and configuration than we planned?" you ask.

"Yes, about two weeks," Jeannie answers. "There is a way to make it go faster."

"How?" you ask.

"Well," she answers, "every time we need a different driver for the NIC, the installers have to either have the disks on hand or log on to the Internet and download the driver from the manufacturer's website. We keep finding the need for more drivers every day, and sometimes the websites are overloaded and we can't access the information. This situation made it impossible for us to complete installation in one area for more than a day."

"It was frustrating," Rex adds. "And when one installer needs a disk that they didn't take with them, then they have to stop and either return to our offices or locate the person with the driver disk."

"Sounds like you're not going in there very prepared," Sean says.

Jeannie frowns. "There's really no way to anticipate what drivers the installers will need for every single computer, and there are so many, they'd have to carry dozens of disks. Right now, we've encountered the need for six different kinds of NIC cards, requiring drivers for every operating system version. What we need is someone to compile all the drivers into one central place where our technicians can access the information from anywhere."

"An FTP site," Rex says.

"FTP?" you ask.

"File transfer protocol," Sean explains. "We'd rent some private space on an Internet server. Our technicians would go to that space to get their drivers. We'd just need to make sure that the server will be a good one, not one that goes down all the time or would allow too few people to access the site at once."

"I'll look into it," you say.

Later, back in your office, you discuss the driver downloading issue with your manager, Carol Girolomi. "We could get an intern to compile all the files," Carol says.

"How about several?" you ask. "We need this done ASAP."

"We might be able to pull a few off other projects," she says. "What about renting server space?"

"I'm looking into the costs," you say.

"Write up your proposal and put it on my desk as soon as possible," Carol says. "I'll get you an answer within twenty-four hours of getting it."

"Thank you," you say. Carol leaves your office, and you immediately turn to your computer. Logging on to the Internet, you find several companies offering FTP sites for many different rates. You choose three to contact, based on price.

The first, an Internet service provider called Univers, offers FTP sites of up to fifty megabytes for $110 per month, with up to fifteen simultaneous users, according to their web site. You call their business sales office and find out that additional space is available for five dollars per megabyte (MB) per month. Their server, according to their representative, has performed at 99.9 percent for the last year.

The second provider is FTPLink, by far the least expensive option at five dollars per MB per month. Their web site claims their server has been online for 95 percent of the previous year. You were unable to get through to a customer service representative—their telephone line was busy, even though you tried several times over a few hours.

The last provider, DataAnywhere, offers space at $150 per month for up to fifty MB, with additional space available in 10-MB units for fifteen dollars per unit per month. After speaking to a customer service representative, you learn that DataAnywhere has two servers, a redundancy that has kept their files available 100 percent of the time. They also claim that the only limit on the number of people using the site is that the same file cannot be downloaded at the same time.

As for your own staffing needs, you speak to the Solutions director of human resources, Joe Brookes. He informs you that he can provide three interns to the project, at a cost to your department of $160 per day.

You believe you have enough information to write a proposal, and begin working on it immediately.

Suggested Next Steps

- Review the guidelines for proposal writing as outlined in this chapter.
- Decide which service provider to use and discuss this decision in your Solution Defense.
- Generate other options to your main proposal and include them in your proposal. Discuss why you believe certain options to be better than others in your Solution Defense.

On-Site Daycare Proposal

Considerations

People who work in Human Resources departments handle a variety of situations relating to a company's employees, such as hiring, firing, and creating or maintaining employee assistance programs. In this situation, you have decided to support the creation of an in-house day care center for the employees' children. As you read this scenario, keep the following questions in mind:

- What is your company's history with day care center proposals?
- What process must you go through to have your proposal considered?
- Who supports a day care center, who opposes it, and what are their reasons for these positions?

You are an assistant human resources manager at Morris, Inc., a marketing consulting firm employing just over 600 people. Morris, Inc. handles major advertising accounts, particularly pharmaceutical firms and financial institutions. As part of your job, you attend staff meetings of the various divisions and report back to your supervisor and human resources director, Kim Tok, with any requests typically handled by human resources. You have only been with Morris, Inc. for eight weeks, and so far you have handled only minor employee complaints, such as mix-ups with health insurance claims and one dispute over assigned parking spaces.

Today, you are meeting with the Jonco group, a team of writers, artists, and marketers handling a multimillion dollar pharmaceutical ad account. You arrive at the meeting room to find almost everyone in attendance. After ten minutes, you wonder why the meeting hasn't started, so you ask Jack Pierson, project manager, if he planned to begin soon.

"We can't start without Kelly," he explains, referring to Kelly Morrison, head of the graphic art team.

"Why isn't she here?" you ask.

"She's been having problems with her children," Jack said. "Seems her day care provider moved away, and she's had a hard time finding someone new."

"She's making arrangements on a day-to-day basis," said Beth Hunt, another graphic artist. "It's really hard to get kids into good day care. I had to wait two months before my daughter got into a good place, and I have to drive a half-hour out of my way every day at that. Why don't we have a day care center here? Other companies do it."

You don't have an answer for her, and you assure the group that you'll bring it up at the next HR meeting. Meanwhile, Kelly arrives, makes her apologies, and the meeting begins.

After the group disperses for lunch, you ask Kelly to stay on for a moment.

"I hear you're having trouble with your child care," you say.

"Yes," Kelly answers. "The woman who had been taking care of my son moved away. It's been difficult finding someone qualified to take over."

"Would you be willing to help plan a company day care center?" you ask.

"I thought the company didn't want to support child care," Kelly says.

"I wasn't aware of that," you say somewhat surprised.

"I sat on a committee last year about this, and we never heard an answer to our proposal," Kelly tells you. "As my contribution, I did a survey of all the parents working here and found out how much they spent on day care and how much work they'd missed because of child care problems."

"I'd like to take a look at that survey," you say.

"I still have the file on my computer," she says. "I'll e-mail it to you."

"Thank you," you answer. "But are you willing to give it another shot?"

"Look," Kelly says, "I really don't have time to spend on something that probably won't happen. You can have the survey, but I doubt many people will want to help plan on-site day care until the management indicates they'll even consider it."

"I understand," you say as Kelly leaves the room.

Later, you knock on Kim's door to give him some reports, and you mention the discussion you'd had with the Jonco group about day care. Kim looks up from his computer, obviously interested. "You know, we tried to discuss this with management last year," he says. "And we didn't even get to make a formal proposal. Some people formed a committee on the subject and sent me a bunch of information, but I could never interest management in considering the idea."

"Do you think they might be interested now?" you ask.

"Perhaps," he replies. "We have two new vice presidents, one of whom recently had a child. And I have to admit that I'm more interested in it now that I have a kid. It would be nice to see Thomas during lunch and to take him to and from work with me."

"So where do we start?" you say. "I have a lead on a year-old survey of how many employees use day care, the amount they spend, and their experiences with it."

"I think I've seen that one," Kim says. "That's a start. Why don't you begin researching the idea—find out what other companies that have on site day care do and whether or not they see benefits. I'll bring this up at the next management meeting and see if we can interest them in receiving a proposal."

You leave Kim's office very hopeful. You check your e-mail and find a pleasant message from Kelly with an attached file. You read the file and take notes on the survey, summarizing the data.

No. of Morris employees with children	144
No. of minor children of employees	212
Under 5	58
5–12	104
13–18	40
Weekly day care expenses	
Range	$50–212
Average	$123
Ave. no. of times late to work because of day care problems in the last year	4
Ave. no. of times left work early because of day care problems	7

You are in the middle of some intensive Internet research on companies with day care programs when Kim knocks on your office door.

"Guess what?" he says obviously happy.

"What?" you ask playing along.

"You have a proposal to write on day care, due next week," he says.

"Fantastic," you reply. "I've already started the research."

"Florence Mather is going to take the lead on the project, if they accept the proposal. She's the one who just had a kid. Oh, and she agreed to provide the space and find renovation funds," Kim says. "They'll even buy the toys and other stuff. If they believe it will improve employee performance. That's your job."

"What about staffing?" you ask.

"We'll have to include that in the proposal. You'll have to find out what qualifications are used to hire day care workers—I'll make out a job description and attach it to the proposal. That way they can approve the description with the proposal, and we can get the hiring process underway."

"I'm on it," you say, and you return to your research.

SUGGESTED NEXT STEPS

- Review the guidelines for writing proposals as outlined in this chapter.
- Conduct research on day care facilities housed within major corporations. Use the information from your research to justify the implementation of on-site day care. Discuss what appeals you are using to convince the company's management in your Solution Defense.
- Generate alternatives to on-site day care and include them in your proposal. In your Solution Defense, discuss what alternatives seem best to you, and why.

Kid's Knob Proposal

CONSIDERATIONS

You have taken on a significant portion of a community project to build a playground. As you read this situation, keep the following questions in mind:

- What are the different organizations involved in this project, and what potential does each one have to assist or defeat the playground proposal?
- What is the process your organization must go through to have your proposal considered?
- Before you write the proposal, what pieces of information must you gather together?

As a student at Western University, you became involved in the town's efforts to provide youth recreation opportunities. Newton, population 45,000, is located in the upper Midwest and is primarily a college town. You sit on the city's Youth Recreation Committee, a group of young people, parents, college students, and city officials working to find cost-effective ways to provide fun things for young people to do. This winter, the committee has taken on the rather large task of building a new playground.

Only two playgrounds exist in Newton, one at each of the town's elementary schools. In the center of Newton is a large park, Centennial Park, the proposed site of the new playground. A local architect has already donated her time and designed the playground, a sprawling collection of towers, swings, poles, and other standards that look somewhat like a castle when put together. According to the architect, the structure will cost about $25,000 in materials and $75,000 in labor to prepare the site, assemble the structure, and complete the finishing work (sanding, testing the equipment, etc.).

The first step in the process, though, is convincing the Newton City Commission that the project is worthwhile and to officially donate the needed space in Centennial Park. The City Commission would also have to approve an additional $5,000 a year in maintenance costs associated with the new playground. Beyond these annual costs, the committee hopes the city will assist with the initial funding of the project.

Funding is the hot topic at tonight's committee meeting.

"They'll never go for it," says Bill Thompson, a former Newton city commissioner. "Even if they got a matching grant, that's still fifty grand out of their pockets. The city just doesn't have that kind of money laying around."

"And there are some important funding issues going on," added Ann Woods, one of the parents on the committee. Ann is also involved in several other city projects, including fundraising for expansion of the Newton Library.

"What if we got a grant for more than just half?" you ask. "Or had volunteers to do the assembly, and convinced local companies to donate their time and equipment?"

Everyone at the table stops talking. You have their attention.

"Who knows someone in the construction business?" you ask.

"I do," Bill answers. "My friend Chris Sanders is a general contractor."

"Will you ask him to consider donating some services?" you say.

"Sure," Bill says. "He just might be willing to come in with a bulldozer to do the leveling."

"Mark," you say, turning to one of the other college students. "You're part of the campus volunteer center. How many people do you think we could round up to help put this thing together?"

"I don't know," Mark answers. "How many do we need?"

Bill turns to Carl Ruskan, who worked with the architect. "Carl?" he asks.

"All the pieces come premade," Carl says. "They're connected by bolts and screws with some nailing required. After that, it's a lot of sanding to make sure there aren't splinters."

"Sounds like something a bunch of college kids could handle," Mark says.

"The more the merrier," Carl replies. "We need at least sixty or seventy people to do this, if we're going to get all the assembly done in one weekend."

"And people to feed these workers," Ann says. "And someone to handle renting portable johns and cleanup."

"This is good," Bill says. "Sounds like we're ready to write ourselves a proposal."

"But we don't know a bunch of things," Ann says. "Like whether or not we can get the people and donations."

"I would write it like this," you offer. "We'll ask for as much money as we've estimated from the city, and we identify some potential sources for matching or full grants, because we can't legally apply for grants ourselves. As alternatives, we list out some of these ideas. While the city is chewing on the proposal, we see how much we can gather together through donations and volunteers. That way, when the commissioners meet to vote, we can tell them just how much we've already managed to accomplish."

The other people around the table murmur their approval.

"Sounds like we have ourselves a proposal writer," Bill says.

"Me?" you say surprised. Everyone laughs.

"Get us a draft by the next meeting," Bill says. "All in favor?"

"Aye!" everyone but you says.

"Opposed?" he asks, looking directly at you. You smile and say nothing.

Suggested Next Steps

- Review the guidelines for writing proposals as outlined in this chapter.
- Research past and present playground construction projects—perhaps some have occurred in your community. Find out how others have used grants, donations, and volunteers to complete these projects. Document your research in your Solution Defense.
- Locate potential sources of funding for this project from local, regional, and national sources. Document your research in your Solution Defense.
- Decide what mix of city and private resources you think are best for this project. Are there reasons to involve the community beyond the issue of funding? In your Solution Defense, explain your reasoning for prioritizing different ways to have the playground built.

Soil Incineration: The Final Proposal

Considerations

This is it—your final document for the soil incineration project. You should review the other soil incineration scenarios and documents before reading about this portion of the project, and keep the following questions in mind:

- What are the positive and negative aspects of the soil incineration project?
- What information do you have that supports the creation of the incinerator, and what information might harm its chances?
- What are the personal and political issues surrounding this project, and what effect do they have on how you will write this proposal?

The time has come to submit an official proposal for the soil incineration project to the County Commission. You and Jack MacGregor have been discussing the project all morning. "Beyond describing the facility so the commission understands its operation, what else are we really concerned about?" you ask Jack.

"The environmental argument, of course," Jack says. "These Protect Our Land people have been screaming *Silent Spring* for the last few months. We have a lot of reassuring to do."

"But they're right about the site's soil being poor for this kind of project, and the location is in a fragile wetlands ecosystem," you say.

Jack glares at you. "You've gone over the containment procedures, haven't you?" Jack asks. "I recall reading that information in your technical description of the facility."

"Yes, I know how we'll be containing the soil and the ash," you reply. "All that information is in the other reports I've created for this project."

"Well, be sure to include all that information," Jack says.

"What about alternative methods?" you ask. "Do you want that included in the report?"

"Yes," Jack says. "But we should be careful to point out how expensive and experimental these methods are," he adds.

You decline to argue with him, because you know that Jack's involvement goes beyond his role as your manager. Jack will be the one to stand in front of the commission and answer technical questions while the incinerator's operators promote the project; Jack will also stand to gain prestige in Donnelly Engineering if the soil incineration plant is approved. If it fails, you have the distinct impression that both you and Jack will experience the consequences.

"One final question," you say. "How much emphasis do you want put on the number of people required to operate the plant?"

"Just keep the technical issues clear and concise," Jack says. "Let me worry about making sure the commissioners know how many jobs this project will create."

After Jack leaves your office, you decide to take a short road trip. You stop at a fast-food drive through and pick up a hamburger. As you head out of town, you wonder if what you're doing is wise.

On the two-track road into the proposed incinerator site, you see several signs, presumably hung by POL supporters. Some are disturbing, others rather clever. "LISTEN TO THE BIRDS SING" one red and white sign proclaims. "WHILE YOU CAN" says the next one. Near the small clearing where the incinerator will be built, there are more signs. "OPEN SEASON ON DONNELLY" one reads. You step out of your car and can't help looking around a bit nervously.

To the west of the clearing, you can see the tops of cedar trees, and you know the wetlands must start just a few hundred feet from where you're standing. Numerous small creeks run out of that swamp, eventually joining up into larger streams with trout and other fish.

The distinctive high call of a crane startles you. The call is answered by several other birds, building into a loud cacophony before abruptly ending. You get back in your vehicle and leave, driving past a few dairy farms on the larger roads leading back to town.

Once in your office, you sit down and stare at the stack of files containing all the other documents you've created for this project. You sigh and begin to go over them one more time before compiling all your information into the proposal.

Suggested Next Steps

- Review the guidelines for writing proposals as outlined in this chapter.
- Discuss your decisions on how to represent certain sensitive issues (particularly the environmental concerns) in your Solution Defense.
- In your Solution Defense, reflect on all the various projects you have completed prior to this final soil incineration proposal. On a personal level, how did your thinking about the project change over time?

Guide to Creating and Working with Visuals

Visuals, or "graphic aids," are very important in technical writing. They allow the technical writer to *show* the reader information, usually in addition to a carefully written description of the item or concept being discussed. Try this simple exercise: write down instructions for people to tie their own shoes. You are only allowed to use words, no drawings. (This will probably take several minutes.) Now have a partner attempt to follow your instructions to the letter. Could your partner tie his or her shoes?

Visuals take the place of someone physically demonstrating a process—like learning to tie your shoes—and they also provide readers with a quick way to make meaning of complex bits of data. As a reader, you probably skip to the pictures, charts, graphs, and tables in a document that you want to read quickly. If you cannot find the information you need or require more explanation, you then backtrack and read the accompanying text. The people reading your technical documents will, more than likely, employ this same kind of "scanning" technique. It's your job as the technical writer to provide not only accurate visuals but also appropriate and sufficient textual information to explain those visuals, thereby enhancing your readers' understanding of the information. This guide provides you with some basic rules of thumb for using visuals in technical documents.

USING VISUALS—WHEN AND HOW

Remember that visuals help your readers organize information and make comparisons not easily done when simply providing information in text. It is not enough to provide numbers, dates, times, and amounts. Your readers must make meaning of this information, otherwise it's all a useless jumble. Take, for example, the following paragraph.

"In the first quarter, each region performed relatively well. Michigan sold 221 units, with Wisconsin and Minnesota turning in 178 and 256 units sold, respectively."

Figure A.1 represents the same information. At a glance, readers are able to make meaning of the data from the paragraph, forming relationships between the regions. After reading the paragraph, most readers take a few more seconds to discern which region posted the greatest average sales than they would by viewing the bar graph.

Another simple method for showing relationships is to create a table, as shown in Figure A.2. The table provides more detailed information, allowing readers to know the *exact* figures involved. Both types of visuals, figures and tables include the following:

- an introduction of the information supplied in the visual
- a caption or title of the visual, numbered consecutively (e.g., Figure 1.1, Figure 1.2, etc.)
- a subtitle labeled "Source" indicating where the statistics were found
- clearly labeled data indicating the type of value represented (region, units)
- a box to separate the visual from the rest of the text

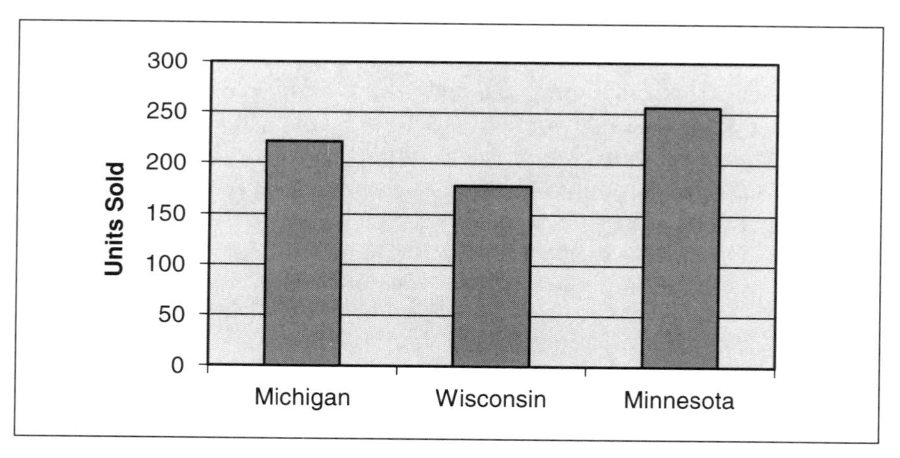

FIGURE A.1 Regional First Quarter Sales

SOURCE: EMI Quarterly Report, 1999.

	Michigan	Wisconsin	Minnesota
Units Sold	221	178	256

FIGURE A.2 Average Quarter Sales by Region

SOURCE: EMI Quarterly Report, 1999.

CHOOSING A VISUAL TYPE

You may choose from a number of different types of diagrams, charts, graphs, and tables. As shown in the previous examples, a table and a bar graph can often adequately represent the same information; however, the table provides specific numbers while the bar graph emphasizes the relationship of one region's sales to that of another. The most general categories of visuals are as follows:

- *Diagrams.* Diagrams are depictions of objects or concepts using drawings, pictures, and callouts (text pointing out aspects of the whole). Figure A.3 is a simple diagram that uses pictures and callouts. You may also take a photograph of an object and add, or overlay, callouts.

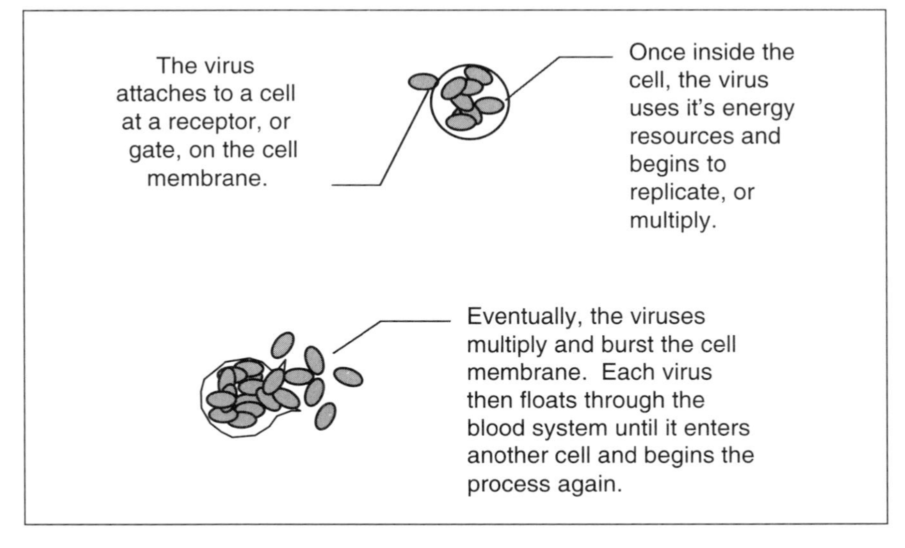

FIGURE A.3 Typical Virus Replication

Customer	Type of Modem	Installation	Monthly Service
Arple, John	3300+	$99	$24.95
Mathews, Celia	5000 AZ	$120	$38.95
Smith, Peter	3300	$88	$24.95
TOTALS	3 Modems	$307	$88.85

FIGURE A.4 Carla's Internet Access Sales, January 1–7

- *Tables*—Tables show specific information and allow readers to compare one set of data with another. Figure A.4 is a simple table that provides both textual and numerical information to emphasize what the author wants to make sure the reader notices.

- *Charts and Graphs*—Charts and graphs organize numerical data in ways that emphasize relationships. Some types of charts and graphs are pie, bar, scatter, plot, line, area, radar, and doughnut. Chose the type of chart or graph that most clearly represents your data and allows the reader to quickly and easily review the information. Many computer programs provide three-dimensional chart and graphs; while these may look interesting, be sure that any graph is, first and foremost, useful to the reader. Figures A.5 and A.6 illustrate a bar graph and a pie chart, respectively. Notice that the bars are labeled clearly and the y-axis (vertical measurement) is labeled so readers can see what the numbers represent (in this example, the number of units

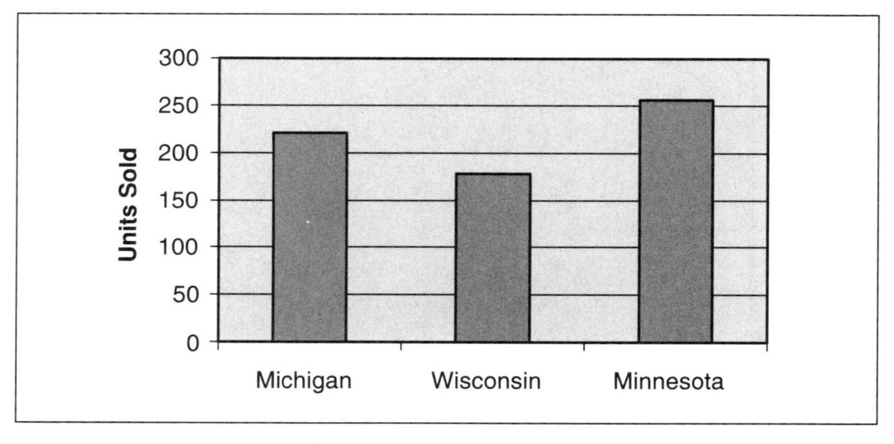

FIGURE A.5 Regional First Quarter Sales

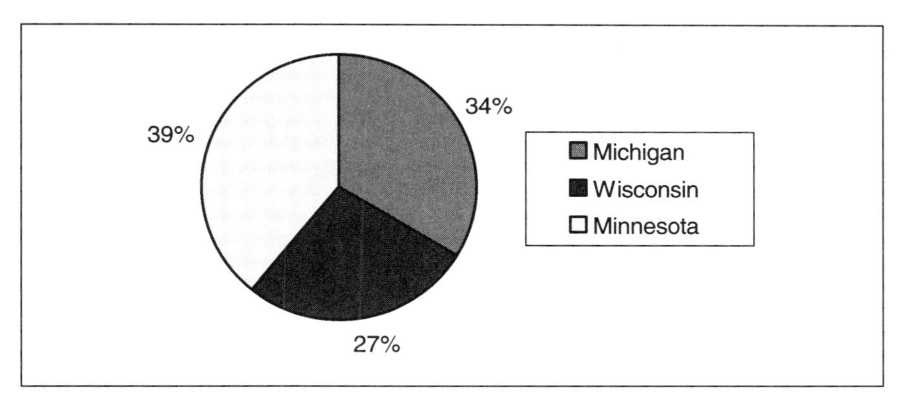

FIGURE A.6 Regional First Quarter Sales

sold). Figure A.6 is a pie chart that shows how extra labeling can help the reader see the differences between items. Without the percentages shown, it might be difficult to tell that Minnesota sold more than Michigan.

- *Pictures and Clip Art.* Pictures are commonly used in technical communication to illustrate a point. You might scan, or digitize, a photograph of a particular piece of machinery, an aerial view of a piece of property, or some other relevant image that further explains a particular idea. Clip art can also be useful when used carefully, particularly common symbols signifying danger or warnings (Figure A.7). Avoid using any image that does not aid the reader in understanding the information presented.

FIGURE A.7 Radioactive Materials Warning

CREATING CLEAR AND ACCURATE VISUALS

Even though you may have picked the right type of visual to use, remember that the way you format the visual also affects the reader's comprehension of the data. Keep the following rules in mind when creating charts, graphs, diagrams, and tables:

- Use a clear font (such as Arial or Times) in a readable size (no smaller than 10 point).
- Limit the use of color to highlighting important data or clearly separating items, such as the columns in a bar graph.
- Create an accurate and succinct title for every visual.
- Use labels to identify kinds of data and measurement units.
- Keep dividing lines far enough away from data so that they do not run into each other. For example, when using a table, make sure you have enough white space within each cell so that the information contained in the cell is clearly separate from the lines that divide the cells.
- Use an appropriate scale to allow meaningful comparisons. Figure A.8 depicts a visual with a compressed scale. Figure A.9 illustrates the same data more appropriately.

Again, keep in mind that your reader may not understand what they need to know by simply scanning the visual. Be sure to introduce properly any graphic, and then explain further the data being represented. An explanation of the data in Figure A.9 might be

> In the traditionally slow months of January and February, Region 4 outperformed most other regions. Region 3 posted significant February sales as well. Overall, Regions 1 and 2 did not move as much product as the noticeably stronger Regions 3 and 4.

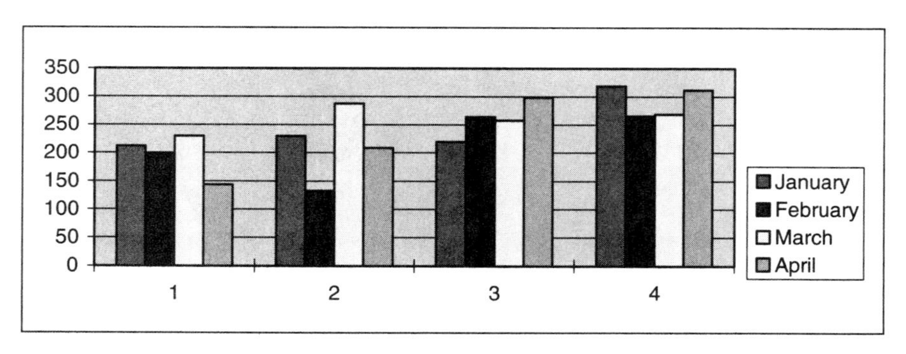

FIGURE A.8 First Quarter Sales by Region

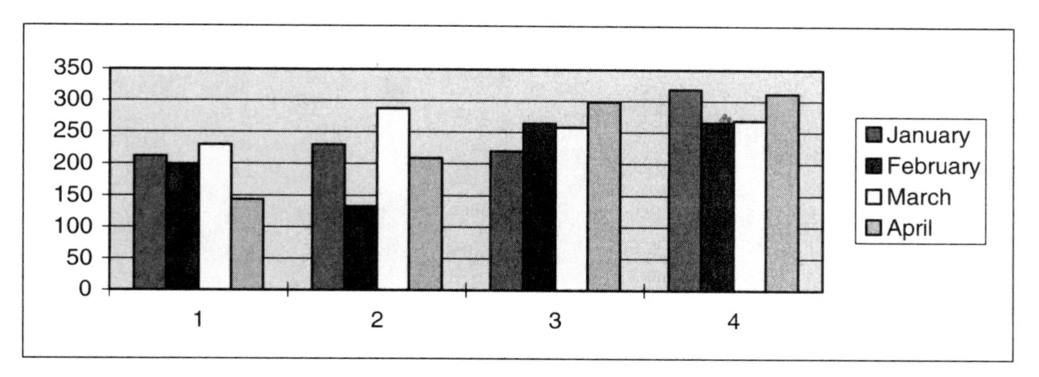

FIGURE A.9 First Quarter Sales by Region

This type of explanation ensures that readers make the kind of comparisons you require for full comprehension of your data. Without such an explanation for Figure A.9, readers would not know that January and February are traditionally slow months and may not have mentally grouped the regions (one and two together, three and four together).

Remember that visuals are purely functional tools in technical documents—they provide your readers with another method of learning. Not everyone can read about a concept or object and understand it; many readers comprehend faster and more completely when you *show* it to them.

Index